AF616121

Host Wall Alterations by Parasitic Fungi

Edited by

Orlando Petrini
Microbiology Institute
Swiss Federal Institute of Technology
ETH-Zentrum
Zurich

Guillemond B. Ouellette
Natural Resources Canada
Forest Service, Quebec Region
Sainte-Foy, Quebec

APS PRESS
The American Phytopathological Society
St. Paul, Minnesota

3 1 JUL 1996

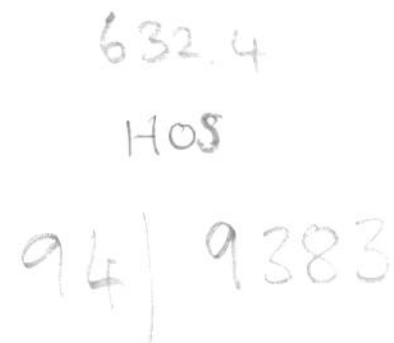

Cover: Interaction between beech (*Fagus sylvatica*) and the anthracnose fungus *Discula umbrinella*. Photo courtesy Olivier Viret

This book has been reproduced directly from computer-generated copy submitted in final form to APS Press by the authors. No editing or proofreading has been done by the Press.

Reference in this publication to a trademark, proprietary product, or company name by personnel of the U.S. Department of Agriculture or anyone else is intended for explicit description only and does not imply approval or recommendation to the exclusion of others that may be suitable.

Library of Congress Catalog Card Number: 94-78283
International Standard Book Number: 0-89054-168-X

© 1994 by The American Phytopathological Society

All rights reserved.
No part of this book may be reproduced in any form, including photocopy, microfilm, information storage and retrieval system, computer database, or software, or by any means, including electronic or mechanical, without written permission from the publisher.

Copyright is not claimed in any portion of this work written by government employees as part of their official duties.

Printed in the United States of America on acid-free paper

The American Phytopathological Society
3340 Pilot Knob Road
St. Paul, Minnesota 55121-2097, USA

TABLE OF CONTENTS

Preface

Infection and colonisation of a host by a fungus include a series of complex processes such as host-recognition, degradation of the host cell wall and formation of specialised infection enzymes and structures by the fungus, as well as of defence structures and compounds by the host. The result is usually the formation of symptoms, although in a number of cases asymptomatic colonisation may also take place.

Upon landing on a potential host, a fungal spore must recognise the host and be activated to successfully start infection. In many cases extracellular sheaths have been believed to serve this scope, and sugars or glycoproteins to be involved in the recognition process. It has now been proved, however, that the extracellular sheaths of spores and hyphae contain active enzymes, such as esterases, that may be involved in the infection process. In any case, the contact between spores or hyphae and the host will most probably elicit defence responses and consequently host-wall alterations.

The morphological and biochemical changes involved in the infection process are the main theme of this book, which results mainly from a symposium held at the Sixth International Congress of Plant Pathology in Montreal, Quebec, Canada, in 1993.

A first chapter (Chamberland) is devoted to a review of electron microscopy methods useful to detect infection and host cell wall alterations by fungi. Fungi are able to form very specialised structures, such as extracellular sheaths or microhyphae, and to induce a number of morphological changes in the host walls. The second chapter, written by Nicole and others, presents the fungal structures and the changes they cause. In a third chapter, the microscopic interactions between fungi and host or non host systems are described by Rioux and Biggs.

Five chapters are devoted to the biochemistry of host cell wall alterations by fungi. Notably, plant cell walls are mainly composed by polysaccharides and by one of the most recalcitrant biological molecules known, lignin. Joseleau and Ruel describe the phenomena underlying wood polysaccharide degradation, and Blanchette presents a brief overview of lignin degradation by fungi.

Kolattukudy and co-workers outline how two extremely important compounds, cutin and suberin, that serve basically to protect the plant from ecological changes and pathogens, are degraded by fungi, and show the role of these substances in the defence mechanisms of plants.

Fungi may also serve as hosts to hyperparasites, and this kind of interaction is important to study biological control. Since the cell wall in fungi is composed mainly by chitin, its degradation is of particular interest, and Manocha and Balasubramanian address the mechanisms underlying chitin degradation. The last chapter dedicated to biochemical aspects of the interaction, written by Sheng and Showalter, deals with expression and roles of plant cell wall structural proteins in fungal infection.

Finally, two symbioses are discussed. Bonfante describes the alterations of host cell walls caused by an economically important group of fungi, the mycorrhizae, whereas Stone and others present the still limited information available on the only recently studied symbioses between endophytes of aerial plant tissues and their hosts.

This book could have not been completed without the help of many individuals. The APS editor, Dr. J. Marois, has been very helpful during all phases of planning and preparation, and his quick, yet very meticulous reviewing of the manuscripts has greatly helped us in the editorial tasks. We should like to thank Drs. J. E. Adaskaveg (Davis, CA, USA), R. A. Blanchette (St. Paul, MN, USA), Th. Boller (Basle, Switzerland), C. Breuil (Vancouver, B.C., Canada), D. L. Brown (Ottawa, ON, Canada), A. J. Buchala (Fribourg, Switzerland), K.-E. L. Eriksson (Athens, GA, USA), R. L. Farrell (Lexington, MA, USA), V. Gianinazzi-Pearson (Dijon, France), R. Hammerschmidt (East Lansing, MI, USA), W. Köller (Geneva, NY, U.S.A.), J. McDermott (Zurich, Switzerland), J. Peberdy (Nottingham, U.K.), M. N. Philipson (Greytown, New Zealand), A. M. Showalter (Athens, OH, USA), Th. Sieber (Zurich, Switzerland), J. K. Stone (Corvallis, OR, U.S.A.), A. Tsuneda (Tottori, Japan), A. G. J. Voragen (Wageningen, The Netherlands), and M. Wingfield (Bloemfontein, R.S.A.), for kindly accepting to review manuscripts. The help by Dr. L. E. Petrini (Comano, Switzerland) has been invaluable in accomplishing all the editorial tasks linked with the preparation of a camera-ready book. Finally the senior editor should like to thank his students for the patience shown during all phases of the preparation of the book.

O. Petrini and G.B. Ouellette,
March 1994

Host Wall Alterations by Parasitic Fungi

GOLD LABELING METHODS FOR THE ULTRASTRUCTURAL LOCALIZATION OF HOST WALL AND PATHOGEN COMPONENTS

Hélène CHAMBERLAND

Department of Biology, Pavillon Vachon, Laval University, Ste-Foy, Québec, Canada G1K 7P4

Much information has been obtained during the recent years on the nature and organization of plant cell walls and of their alterations. Numerous staining techniques have been applied at the light microscope (LM) level to show that suberin, callose, lignin and other phenolic compounds are deposited in plant cell walls as a response to pathogenic or abiotic stress. LM and transmission electron microscopy (TEM) staining methods, including the PATAG staining technique (487), have provided important information on wall changes and on plant-fungal interfaces (84, 112, 166, 221, 298, 410, 445, 478).

The introduction of colloidal gold as an electron-opaque marker has provided a means to localize ultrastructurally specific molecules (192, 209, 232). Due to their high electron opacity and small size, gold particles allow an exact localization of labeled sites. Monodisperse particles ranging in diameter size from 2 to 150 nm can be easily prepared and remain stable for months (209). Probes of 1.4 nm in diameter are now commercially available that attach more easily to the target molecules (200). Being negatively charged, gold particles can be complexed by non-covalent adsorption with various kinds of molecules, and are therefore excellent markers for TEM and scanning electron microscopy (SEM). In addition, the high opacity of gold particles makes possible relative quantitation of the labeling. The advantages of colloidal gold labeling have allowed new information to be obtained about the nature and structure of walls of plants and fungi. This approach has led to a better understanding of how wall components are altered in a plant/parasite interaction.

TISSUE PROCESSING

Post-embedding labeling with gold probes is the most popular technique applied in ultrastructural studies. Both fixation and embedding procedures are of importance in cytochemistry and immunocytochemistry. They can alter the conformation of macromolecules, thus limiting their affinity properties (32). In any cyto- or immunocytochemical work, it is therefore necessary to first establish conditions for tissue processing to obtain optimal labeling in cellular structures.

Several studies on the ultrastructural localization of wall sugars were carried out on tissues fixed with glutaraldehyde, formaldehyde, or a mixture of these two aldehydes and further submitted to post-fixation with osmium tetroxide. Cellulose, polygalacturonic acids and pectin have been localized by gold labeling in double-fixed tissues (38, 39, 77, 98, 342, 345). These results may be explained by the fact that osmium tetroxide generally does not interact with most of the pentose or hexose sugars or their polymers (213). Hemicellulosic saccharides such as xyloglucan, rhamnogalacturonan and arabinogalactan have been detected in tissues fixed with aldehydes only (352, 420, 503, 504).

Proteins are generally more sensitive to tissue processing than saccharides. As they are detected through antibodies raised against purified peptides or proteins, all conformational changes caused by tissue processing will result in a substantial reduction in antibody recognition. Mild aldehyde fixations are usually most appropriate for immunocytochemistry. Glutaraldehyde cross-links proteins better than formaldehyde does but it may diminish immunoreactivity. Therefore, formaldehyde is preferred to glutaraldehyde even though the ultrastructural preservation of tissues fixed by formaldehyde is not as good as with glutaraldehyde. These fixatives can nevertheless introduce free aldehyde radicals into the specimen that may cause non-specific labeling. It is therefore recommended to carefully wash samples after fixation with aldehyde blocking substances such as glycine, lysine or NH_4Cl after the fixation step to eliminate free aldehyde groups (31). Sodium borohydrate (1 mg/ml) may also be used for that purpose (512).

Postfixation with osmium tetroxide reduces antigenicity of proteins and is generally not used for their localization. However, antigenicity can be restored by treating sections with oxidants such as saturated sodium metaperiodate which can be applied alone or in combination with hydrochloric acid (33). Such treatments are often effective, but they may produce other types of alterations (83).

Although mild fixation with aldehydes are commonly used in immunocytochemical work, some antigens may not withstand such treatment. Cryofixation and other cryoprocessing techniques better preserve immunoreactivity. Unfortunately, cryofixation is not easily applicable to higher plants. Rapid freezing is difficult to achieve with these tissues as their cells are thick-walled. In addition, the large watery vacuoles present in many plant tissues make rapid cryofixation nearly impossible; therefore, the use of cryoprotectants before cryofixation is recommended. Best results have been obtained for plant and fungal tissues that have a low water contents such as seeds, pine needles, or dormant spores. Nevertheless, many specimens have been frozen in their native state (411).

Cryoultramicrotomy is not frequently used in immunocytochemistry because of the difficulties encountered and because it needs expensive instrumentation and good expertise. Thus, this technology is not frequently used for plant specimens. It has been applied to localize storage proteins in bean cotyledons (195).

Cryofixation methods involving high pressure freezing have recently been developed. In contrast to other cryofixation techniques, high pressure freezing permits the preservation of relatively thick plant samples (477). In immunocytochemical studies on *Brassica,* this approach has improved both ultrastructural preservation of stigma papillae and the localization of specific glycoconjugates (250). Recently, this method has been applied to elm wood samples inoculated with *Ophiostoma ulmi.* Although wall detachments occasionally occurred, the ultrastructural preservation of fungal cells was generally improved: plasmalemma, outer mitochondrial membranes, nuclear envelopes and tonoplasts exhibited smooth profiles when compared with chemically fixed material (Figs. 1, 2). In addition, microtubules and multivesicular bodies were more easily discernable. An efficient gold labeling was seen in samples processed by high pressure, even though they had been osmicated. Labeling for DNA of *O. ulmi* cells in elm wood sections was effective and the cell structures well preserved (Fig. 3). Gold particles were mostly associated with nuclei but were also found over mitochodria and unexpectedly, but in accordance with another study (439), over portions of the cytoplasm.

In cyto- and immunocytochemical studies embedding can diminish greatly tissue permeability leading to a reduction of labeling. Comparative studies have been carried out in animal tissues which demonstrated the importance of tissue processing for immunocytochemistry. Epoxy resins such as Spurr or Epon/Araldite are known to be highly hydrophobic and to affect especially antigenic preservation. Thus, for immunocytochemical studies, polar acrylic resins which are hydrophilic in nature are more appropriate (32, 339). Numerous investigations carried out with resins such as Lowicryl K4M or HM20, LR White or LR Gold, yielded satisfactory labeling of tissues. Lowicryl, which polymerizes under ultraviolet light and can be used at low temperature, is particularly efficient in preserving immunoreactivity. New resins suitable for embedding at temperatures as low as -80°C (Lowicryl K11M, HM23) or with sectioning

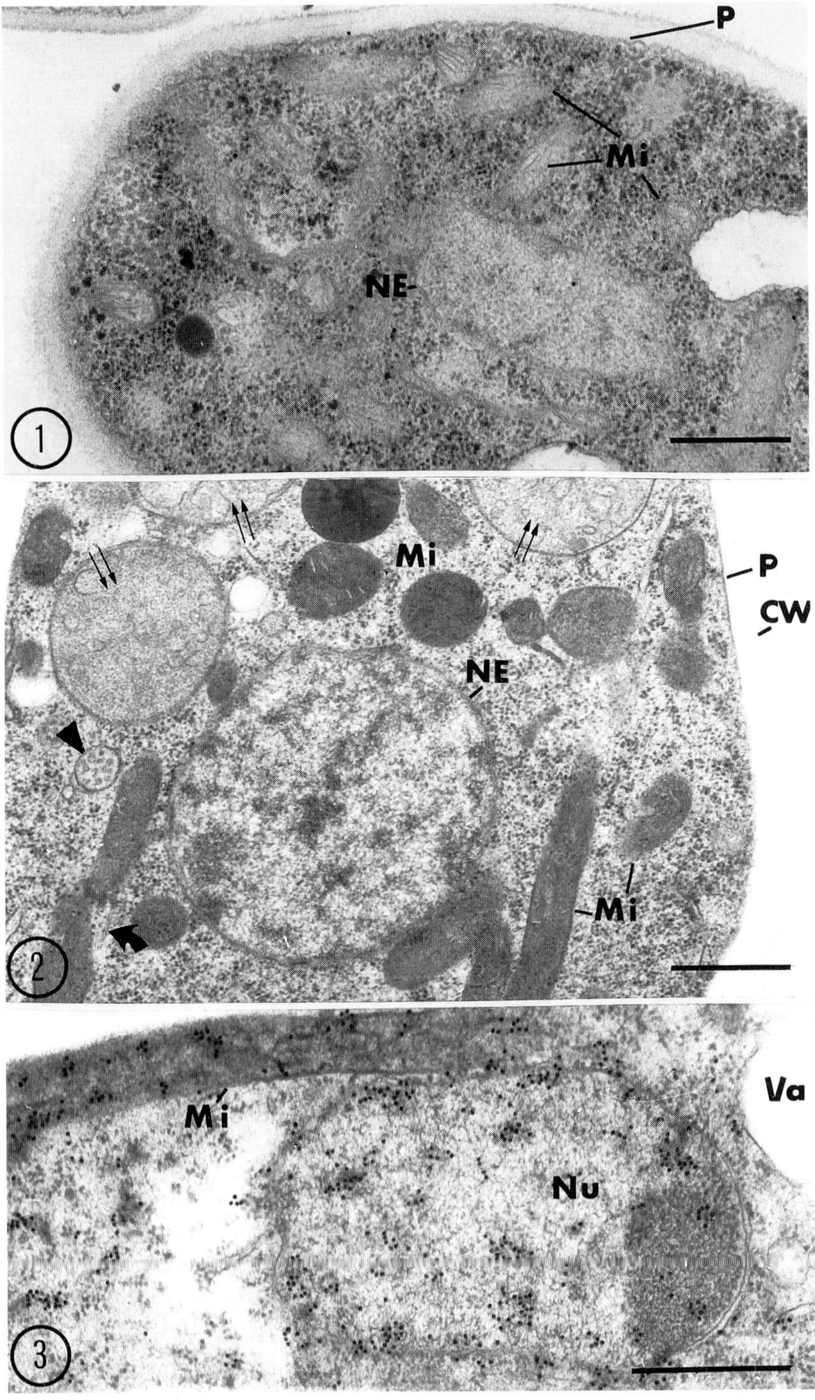
P
Mi
NE
1
Mi
P
CW
NE
Mi
2
Va
Mi
Nu
3

characteristics that favor labeling (Unicryl) are now available (227, 436). Some proteins have nevertheless been localized in samples embedded in hydrophobic media. This is the case for pectin esterase, which has been detected in tomato root tissues embedded in Epon (100).

LECTIN- AND ENZYME-GOLD LABELING

Owing to their peripheral location, sugar molecules are the first cellular components to be involved in host-pathogen interactions. Numerous studies carried out with lectins or enzymes conjugated to colloidal gold have helped to localize various wall components. Lectins are proteins or glycoproteins which specifically bind sugar residues (188). When complexed to colloidal gold they constitute reliable probes for the ultrastructural localization of saccharides (417). Some lectins, including wheat germ (WGA) and *Limax flavus* agglutinins, however, are not successfully adsorbed to colloidal gold and have to be conjugated with colloidal gold using an indirect procedure, in which a glycoprotein specific for the lectin is conjugated to colloidal gold and used as a second-step reagent. Lectin-gold labeling has been used to study cell wall composition (34). In higher plants, galactose, N-acetyl-glucosamine and polygalacturonic acids have been localized in cell walls by means of the *Ricinus communis* agglutinin, WGA, and *Aplysia depilans* lectin (Table 1). By allowing the specific localization of sugars, these probes were useful in demonstrating wall degradation or new wall appositions in host-pathogen interactions and symbiotic relationships (Table 1), or mycoparasitism (Jabaji-Hare, personal communication). For example,

---

Figs. 1–3. 1. *Ophiostoma ulmi* cells fixed with glutaraldehyde, postfixed with osmium tetroxide and embedded in Epon. The plasmalemma (P), nuclear envelope (NE) and outer membranes of mitochondria (Mi) exhibit an undulated profile. **2–3.** *O. ulmi* cells fixed by high pressure freezing technique, freeze substituted in acetone containing 2% osmium tetroxide and embedded in Epon. **2.** In contrast to Fig. 1, membranes appear smooth. A microtubule is discernable (arrow) and large organelles filled with a greyish material and containing numerous membrane systems are present (double arrows). A multivesicular body can be seen (arrowhead). CW: cell wall. **3.** After incubation of sections with a monoclonal anti-DNA antibody (first step) and a gold-conjugated anti-mouse antibody (second step), labeling is present over dense chromatin masses of the nucleus (Nu) and also over a mitochondrion (Mi). Few gold particles are also seen over cytoplasmic opaque material, whereas in other cellular compartments including a vacuole (Va), no gold particles are noted. Bar: 0.5 μm.

the use of a lectin specific for polygalacturonic acids demonstrated that host wall degradation was associated with alteration of pectic material in tomato leaf tissues infected by *Clavibacter michiganense* (36). Other labeling experiments showed that pectolytic enzymes produced by *Fusarium oxysporum* f. sp. *radicis-lycopersici* in tomato root tissues act through localized degradation (38), thus confirming conventional TEM observations (105). Furthermore, accumulation of pectic substances in altered phloem tissues, deposition of polygalacturonic acids and galactose as wall appositions or papillae could be demonstrated by lectin-gold labeling (38, 98).

Although several lectins are known, not all cell wall components can be specifically recognized by them. This is one of the reasons why enzymes have been linked to colloidal gold to detect polysaccharides. First used to localize nucleic acids (29), enzyme-gold labeling has been employed to detect other substrates. This approach provides a high specificity towards a substrate (30). In addition, enzymes can be easily conjugated to colloidal gold and labeling experiments can be rapidly carried out (one-step labeling). However, purity of the enzyme preparation, isoelectric point of the enzyme and its molecular weight, and the pH of optimal activity have to be taken into account when preparing enzyme-gold complexes. In addition, the size of the gold particles may be important to obtain a stable enzyme-gold complex. Best results were obtained for xylanase and mannanase when small gold particles (4-9 nm) were used, being more in proportion with the size of the enzymes (420). Improved labeling is also sometimes obtained when enzymes are heat-deactivated prior to preparation of gold complexes. For example, two xyloglucan hydrolases thus treated gave strong and specific labeling in contrast with unheated preparations (504). Without heat pretreatment, the enzyme-gold conjugates may either modify the target substrate or digest it partly and dissociate from the sections, thus leading to a weak labeling. So far, several enzyme-gold complexes have been used for the ultrastructural localization of cellulose and hemicellulosic components (Table 1). Recently, a purified commercial β-1,3 glucanase was complexed to colloidal gold and used to localize β-1,3 glucans either in plant tissues (Fig. 4) or in the extracellular sheaths of *Ophiostoma ulmi* cells (Fig. 5). A glucanase from another source was also employed to detect β–1,3 glucans in papillae formed in infected tobacco roots (37).

Lectin- and enzyme-gold technique are powerful tools to study cell surfaces and have been used in many studies to show that sugar molecules of cell walls are involved in the establishment of cellular interactions (Table 1).

Table 1. List of some higher plant cell wall molecules detected by gold labeling techniques. l: lectin; m: monoclonal antibody; p: polyclonal antibody.

Molecules	Probes	References
cellulose	exoglucanase	39
	cellobiohydrolase & endoglucanase	64
pectic compounds		
polygalacturonic acids	*Aplysia depilans* (l)	41
pectin	antibody (m)	259
rhamnogalacturonan	antibody (p)	333
arabinogalactan	antibody (p)	352
hemicelluloses		
xylan	xylanase	503
	antibody (p)	352
xyloglucan	antibody (p)	333
	endoglucanase & galactosidase	504
callose	antibody (p)	352
	antibody (m)	319
	β-1,3 glucanase	37
galactose	*Ricinus communis* (l)	40
N-acetyl-glucosamine	WGA (l)	99
HRGP*	antibody (p)	44, 469
GRP*	antibody (p)	427
PRP*	antibody (p)	326
chitinase & glucanase	antibody (p)	42, 311
pectin esterase	antibody (p)	100

*HRGP: hydroxyproline-rich glycoprotein; GRP: glycine-rich protein; PRP: proline-rich protein

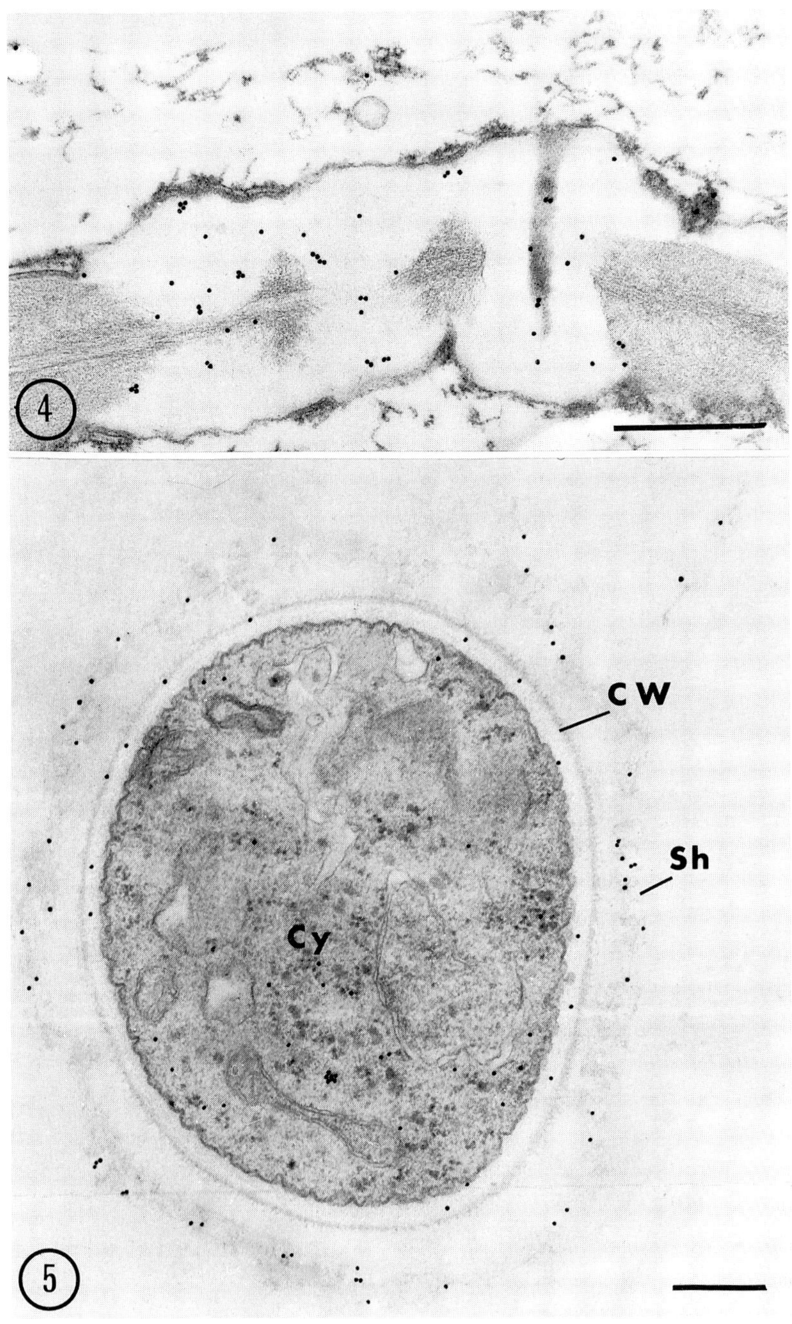
4
5
CW
Sh
Cy

IMMUNOGOLD LABELING

Most immunocytochemical studies on the ultrastructural localization of plant molecules have been carried out by indirect labeling procedures using specific antibodies in a first step and protein A-gold or gold-conjugated secondary antibodies in a second step (500). Polyclonal and monoclonal antibodies can be produced (133, 173). Raised in rabbits immunized with a purified antigen, polyclonal antibodies (PAbs) contain a population of antibodies against a wide range of determinants of that antigen. In contrast, monoclonal antibodies (MAbs) are homogeneous and recognize a single determinant of the antigen. By their ability to recognize a single epitope, MAbs present few of the non-specific binding problems observed when using PAbs, particularly those encountered when PAbs are raised against glycoproteins which may have common oligosaccharide side-chains (500). Such specificity, however, may also be problematic if the single determinant recognized by a monoclonal antibody is denaturated or altered during fixation and embedding. In contrast to PAbs, the production of MAbs is expensive and time-consuming. Once produced, however, unlimited amounts of antibody can be obtained. Specificity of the labeling obtained with PAbs as well as with MAbs should always be verified through proper control experiments (223).

Polyclonal and monoclonal antibodies have been used in conjunction with colloidal gold for the ultrastructural localization of wall proteins and of esterified and unesterified pectin, arabinogalactan, rhamnogalacturonan, xylan and callose (Table 1). In some cases the use of specific antibodies confirmed results obtained by means of enzyme or lectin-gold labeling. For instance, a positive reaction for pectic substances was detected in the cell walls of *Fusarium oxysporum* f. sp. *dianthi* colonizing carnation roots when a specific monoclonal antibody (JIM5) was used (Fig. 6). This does not necessarily indicate that pectic substances are constituents of cell walls of this pathogen, although the possible presence of such compounds in fungal walls has been suggested in previous studies (35, 45). Immunogold labeling also led to the detection of hydroxyproline-rich

---

Figs. 4, 5. Sections incubated with a β,1-3 glucanase-gold complex. **4**. Sample of *Hevea brasiliensis* stem phloem fixed with glutaraldehyde, postfixed with osmium tetroxide and embedded in Quetol. Gold particles are present over the more translucent wall areas of the sieve plate. **5**. *O. ulmi* cells fixed with glutaraldehyde, postfixed with osmium tetroxide and embedded in Epon. Gold labeling is seen over the extracellular sheath (Sh). Gold particles are irregularly distributed over the cell wall (CW) and cytoplasm (Cy). Bar: 0.5 μm.

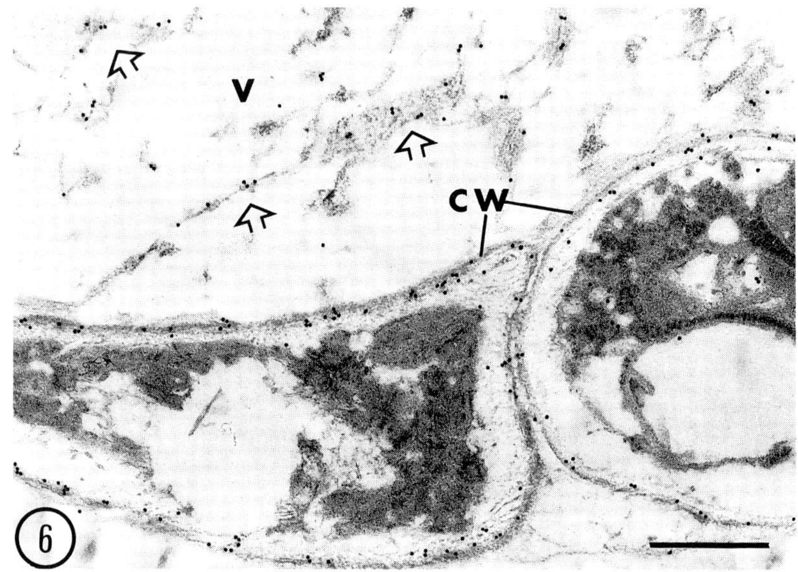

Fig. 6. *Fusarium oxysporum* f. sp. *dianthi* cells in carnation stem tissues fixed with glutaraldehyde and embedded in Epon. Sections were incubated with a monoclonal anti-pectin antibody (JIM5) followed by a gold-conjugated anti-rat antibody. Numerous gold particles are observed over the fungal cell walls. Few gold particles are also present over the fibrillar-like material in the vessel lumen (arrows). Bar: 0.5 µm. (Courtesy G.B. Ouellette).

glycoproteins (HRGPs), glycine-rich proteins (GRPs), proline-rich proteins (PRPs), pectin esterases, chitinases and glucanases in the cell walls (Table 1). Besides plant molecules, fungal components including β-1,3-glucans, fimbriae, enzymes (cellulases, hemicellulases, peroxidases, laccases) or toxins have also been localized either within host walls or within extracellular fungal sheaths by the immunogold approach (43, 47, 126, 161, 245, 344, 422).

FUTURE DEVELOPMENTS

The introduction of more hydrophilic embedding media, new instrumentation for cryo-methods, and continuing progress in the development of microscopic techniques, coupled with advances in biochemical methodologies, have contributed to advances in cytochemistry and immunocytochemistry.

Confocal scanning laser microscopy (CSLM) is a new promising technology that may complement immunocytochemical TEM observations. CSLM can deliver series of sharp optical slices of thick objects, thus providing the acquisition of new data on cell structure and function (366, 521). Gold-labeled probes can also be used as reflection stains (433). CLSM will surely prove of great value in the study of host-parasite interactions. The possibility of studying the 3D distribution of macromolecules in whole cells or thick sections will facilitate investigations on recognition mechanisms or defense reactions.

In situ hybridization (ISH), which is also currently applied with gold-labeled probes, offers the possibility of localizing the genes and transcripts that code for specific proteins. First used in light microscopy (174), this technology is now applied at the TEM level (315). ISH offers the opportunity of studying the regulation of gene expression either temporally or spatially. In plant pathology, RNA probes have been used, for instance, to localize specific mRNAs in infected parsley leaves (437).

In the near future, the integration of molecular and histological methodologies will promote rapid advances in plant pathology and provide a better understanding of the various mechanisms involved in plant-pathogen interactions.

ACKNOWLEDGMENTS

I am grateful to Dr. J. G. Lafontaine (Biology Department, Laval University, (Canada), for critical reading the manuscript and to Mrs. M. Simard and N. Lecours and Mr C. Moffet for excellent technical assistance. I am also grateful to Dr. K. Roberts (John Innes Inst., England) who kindly provided anti-pectin monoclonal antibodies and to Dr. M. Müller (Laboratory for Electron Microscopy I, Zurich, Switzerland) for his contribution to preparation of cryo-fixed samples by the high pressure freezing method.

FINE MORPHOLOGY OF FUNGAL STRUCTURES INVOLVED IN HOST WALL ALTERATION

Michel NICOLE[1], Katia RUEL[2] and Guillemond B. OUELLETTE[3]

[1]ORSTOM, 911, avenue d'Agropolis, B.P. 5045, F–34032 Montpellier, France
[2]Centre de Recherches sur les Macromolécules Végétales, CNRS, Grenoble cedex, France
[3]Natural Resources Canada, Forest Service, Quebec Region, 1055 rue du P.E.P.S., Sainte-Foy, Quebec, Canada G1V 4C7

Alteration of host cell walls (HCW) by fungi is probably a key event in pathogenesis, by which pathogens penetrate plant cell walls and secrete specific wall-degrading enzymes. Numerous biochemical, histological, and cytochemical studies of host-fungi interface have revealed that pathogenic fungi produce a wide range of penetrating structures before plant infection and HCW degradation.

This review focuses on fungal structures associated with HCW penetration and degradation in different plant-fungus associations. Particular attention will be given to the morphology and roles played by extracellular fungal sheaths in cell wall alteration and penetration. Unusual structures that may contribute to HCW degradation will also be discussed.

PARTICULAR CELLULAR STRUCTURES

Pathogenic fungi can produce a wide array of cellular structures specialized in cell wall penetration, such as appressoria, haustoria, or microhyphae. Alteration of plant surfaces results from mechanical pressure and enzymatic activites (332). Involvement of enzymes, i.e. cutinase and pectinase (274) and cellulase (281) in the penetration process has been suggested and partly demonstrated for appressorial fungi, although immunocytochemical evidence for the presence of these enzymes is lacking.

Appressoria

For several fungi such as rusts (228, 229), *Colletotrichum* species (332), endophytes (Stone et al., this volume) and vesicular-arbuscular mycorrhizae (Bonfante, this volume), the interaction with the host surface results in the formation of appressoria.

Appressoria were first described in 1883 as spore-like organs formed on germ tubes of *Colletotrichum* species (154). They are usually described as forms that provide anchorage for infection pegs or germ tubes, and are able to overcome the host's physical barriers such as the cuticle layer (154). Their formation starts when growth of germ tubes has stopped. The fungal cell wall, cytoskeletal microtubules, and apical vesicles are probably involved in differentiation of appressoria (229). Appressoria are often associated with stomatal penetration, but they have also been described in intact plant surface penetration (321, 327).

Initiation and maturation of appressoria are controlled by several factors, including temperature, plant exudates, and the fungal genotype (154). During appressorium formation, chitin of the fungal wall was not detected, likely resulting from glucan overlay (322). Penetration of host surfaces by appressoria requires signaling that may result in tight adherence to the substrate (229, 340). The extracellular matrix of these structures contains specific proteins involved in recognition (156, 322) and initiation of plant cell wall degradation (228, 317). Infection pegs, or appressorial cones, arise from the appressorium and subsequently penetrate the HCW (327).

Figs. 1–5. Microhypha development in HCW. Double fixation with glutaraldehyde and OsO_4; Fig. 1: glutaraldehyde alone. **1**. *Phellinus noxius* in sterilized wood sections. Labeling for cellulose with gold-labeled exoglucanase. The fungal cell plasmalemma is in close contact with HCW, and from or through it radiate arrays of fibrillar strands extend into HCW (small arrows). Bar: 0.2 µm. **2A**, **2B**. LM observation of microhyphae (arrows) grown through or between cell walls, at times continuous with similar structures in cell lumina. Bars: 5 µm. **3**. *Phialophora mutabilis* in copper-chromium-arsenic treated birch. Fungal wall lacking at times (arrowhead) or unperceptible. The cavity margin is marked by a halo of opaque matter (arrows); fibrillar strands or tubular-like structures extend from the fungal cell wall or contents through the cavity to the halo (small arrows). Bar: 0.2 µm (Courtesy of Drs. Hale and Eaton). **4-5**. *Ophiostoma ulmi* in sterilized elm wood sections. HCW degradation around microhypha (arrows) in gelatinous fiber wall layer (Fig. **4A**) and in secondary walls of vessel elements or of parenchyma cells (Figs. **4B**, **4C**). HCW portion between the opaque halos and the fungal cell (small arrows) has apparently not yet been degraded in 4C, whereas its alteration is almost complete in 4B. In both, apparent absence of wall around the fungal cells. Septate microhypha in a fiber cell lumen. **4A**, **5**. Labeling with polyclonal antibodies to fimbriae, predominating over the area corresponding to the fungus plasmalemma (or its invagination) or inner portion of the wall; rows of gold particles in close contact with HCW indicate lack of a wall or presence of a very thin one (Fig. **4A**, arrowhead). Scale bars: 4A, 0.5 µm. 4B, 0.3 µm. 4C, 0.4 µm. In Fig. 5 note much less pronounced labeling in the septate microhypha (arrow). Bar: 0.5 µm.

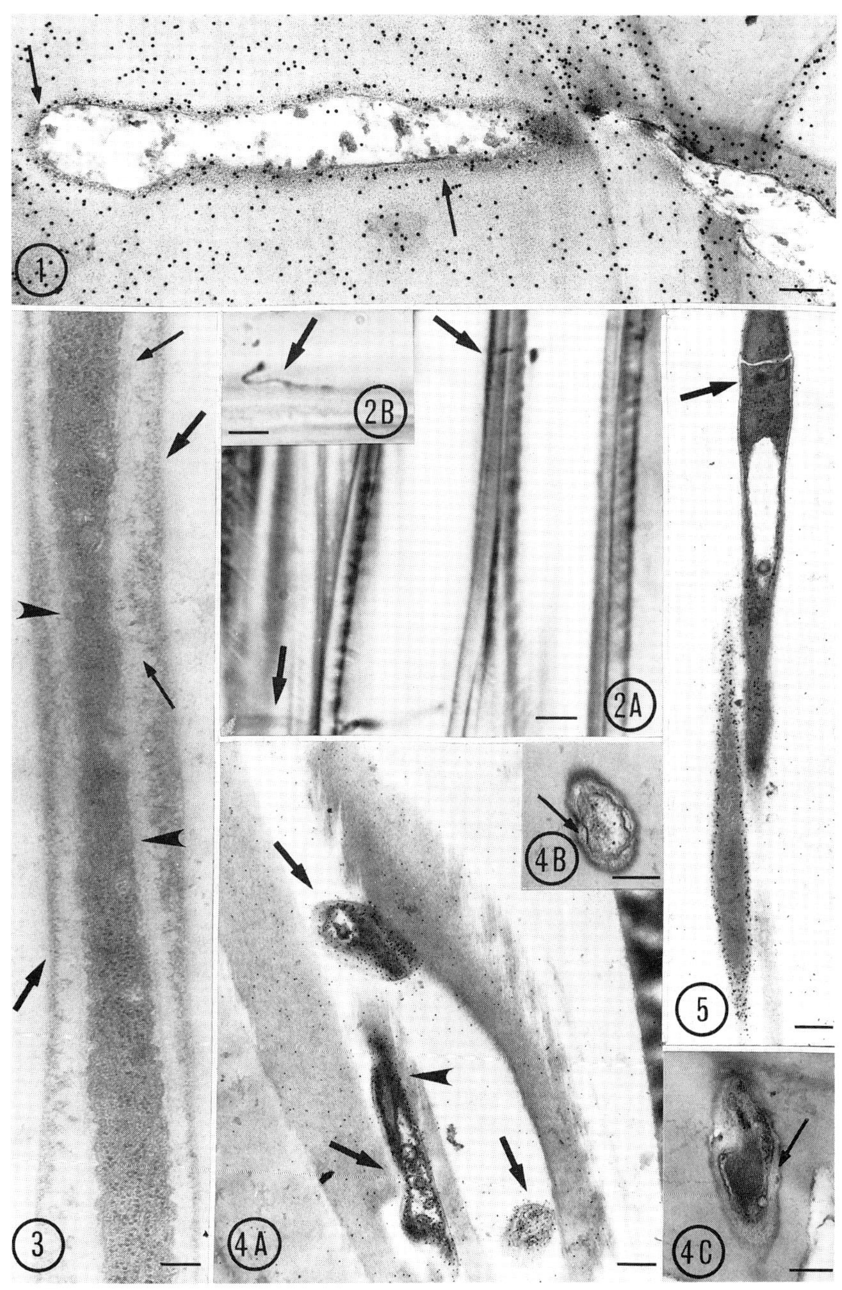
1
2B
2A
3
4A
4B
5
4C

Haustoria

After contact with the host is established, the pathogen develops penetration pegs, vesicles and inter- or intracellular mycelium, as well as haustoria, which start colonisation of host tissues. Haustoria are produced by fungi such as rusts, powdery mildews, downy mildews, and the Exobasidiales. Previous ultrastructural work (149) distinguished four types of haustoria based mainly on the morphological association between the haustorium and the host cell. For instance, in *Phytophthora* species they range from simple, button-shaped to digit-like structures. Haustoria of *Exobasidium* spp. are short and lobed while those of *Cercosporidium* are highly branched (327).

Haustoria arise either directly from penetration pegs or from intercellular hyphae initiated by substomatal vesicles (210, 296). Elongation of the haustorium neck results in the emergence of an haustorium surrounded by extrahaustorial electron-dense matrix and membrane (7, 89). HCW penetration is associated with the formation of an electron-dense penetration matrix which is thought to become the extrahaustorial matrix. High fungal metabolic activities that have been characterized during haustorium formation are probably related to the digestion of HCW at penetration sites.

During penetration, the host plasma membrane is invaginated (149). The extrahaustorial membrane is continuous with the host plasmalemma and strongly adheres to the haustorium. Although the extrahaustorial membrane is presumed to serve as a boundary between the host cytoplasm and the fungus, it may be involved in active exchanges (322), mainly to provide nutrients to the pathogen.

Haustorial wall components are probably involved in recognition mechanisms between avirulent races of stem rust and resistant wheat

---

Fig. 6. *Ophiostoma ulmi* in xylem tissues of naturally diseased trees (1978) sampled the following spring, when disease recurrence was noted. **6A**, **6C**. Opaque halos in HCW in apparent continuity with outside wall layer of the fungal cell (small arrow). **6B**. Large cavity formed in unmodified finer wall layer with only a portion of the halo remaining (lower part). Scale bars: 6A, 0.4 µm; 6B, 0.2 µm; 6C, 0.3 µm.

Figs. 7-8. Microhyphae (arrows) in walls of proto- and metaxylem vessel elements in a susceptible carnation infected by *Fusarium oxysporum* f. sp. *dianthi*. Labeling with a colloidal gold-complexed WGA lectin for chitin revealed a very thin, but apparently discontinuous wall around these hyphae (Fig. **7**, small arrows). Note strong labeling of the secondary thickenings (ST). The irregular microhyphae in Fig. **8** (small arrow) are about 0.1 µm in diameter. Scale bars: 0.5 µm.

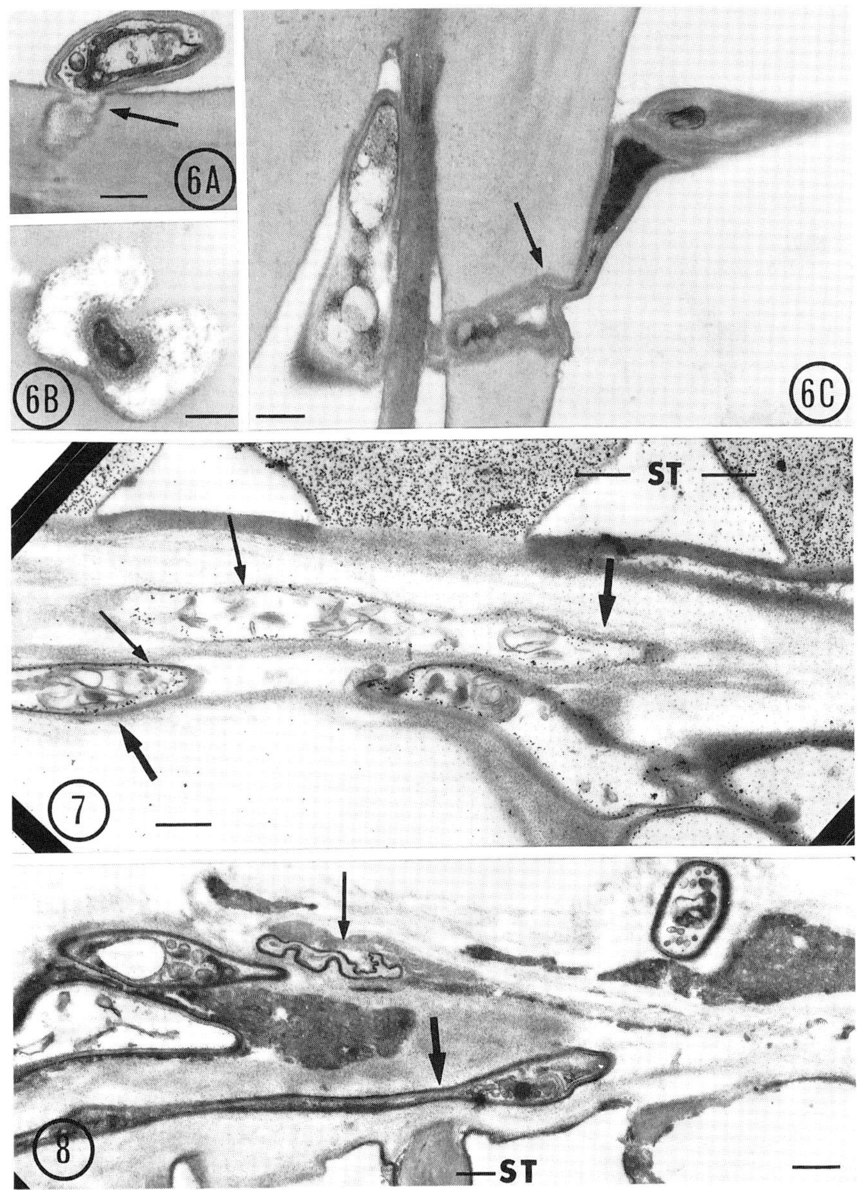
6A
6B
6C
ST
7
8
ST

(113). Recent work indicates that polysaccharides and glycoproteins are present in the haustorium wall and extrahaustorial matrix. A linked-glucose or mannose was found within the *Puccinia graminis* extracellular sheath (ES) (113) while β-1,3 glucans and β-1,6 linkages were identified in ES of *P. sorghi* (106). Haustorium-like structures found in endophytes (Stone et al., this volume) and root-rot fungi (347) do not display the same specialized organization found in true haustoria.

Microhyphae

Fungi may differentiate microhyphae (MH), also called fine hyphae, that can extend for appreciable distances into host walls and cause their breakdown and/or rupture. Microhyphae were first reported in soft rot and blue stain fungi (292), and *Ophiostoma ulmi* (359), and then in Ascomycotina (including Deuteromycotina) and Basidiomycotina (123, 290, 336, 369, 370, 528). *Phellinus noxius*, a decay fungus of tropical trees, has also been shown to produce microhyphae in its attack of wood cell walls (Fig. 1).

The association of MH with cavity formation by soft rot fungi in the S2 cell wall layers can be seen in the light microscope (96); the presence of MH between or within cell walls is generally not evident in LM (Figs. 2A, 2B). Transmission electron microscope (TEM) studies have confirmed the formation of MH and give a better insight into their structural organization and mode of action.

Leightley and Eaton (290) and Hale and Eaton (201, 202, 203),

---

Figs. 9-12. Extracellular sheaths (ES) in four fungi. TEM observations except Fig. 11. **9**. Thick ES around or between *Gremmeniella abietina* cells, strongly labeled for RNA with gold-labeled RNAse. Scale bar: 0.2 µm. Courtesy of Dr. Nicole Benhamou. **10A**. Loose ES around cells of *Ophiostoma ulmi* in sterilized American elm wood fixed with high pressure freezing. **10B**. Apparent fibrils or strands radiating from or through fungal walls into HCW (small arrows). Scale bars: 0.2 µm. **11**. SEM observations of ES coating hyphae (h) of *Phellinus noxius* developing in sterilized birch sections. Note close adherence of ES to host wall surface. Scale bar: 20 µm. **12A–12C**. *Fusarium oxysporum* f. sp. *dianthi* in infected susceptible carnation, labeled for chitin (**12A**) or cellulose (**12B**, **12C**). **12A**. ES-like layer around fungal cells, containing tubular-like structures (small arrow) that extend within the host wall of the vessel element. **12B**. Masses or bands of opaque matter, similar to and continuous with the outside fungal wall layer (small arrows), that extend within HCW (h: hyphae). **12C**. A band of similar matter along a vessel wall, comprising stacks of fibrillar strands, that extend into HCW (small arrow). Scale bars: A, B, 0.4 µm; C, 0.3 µm.

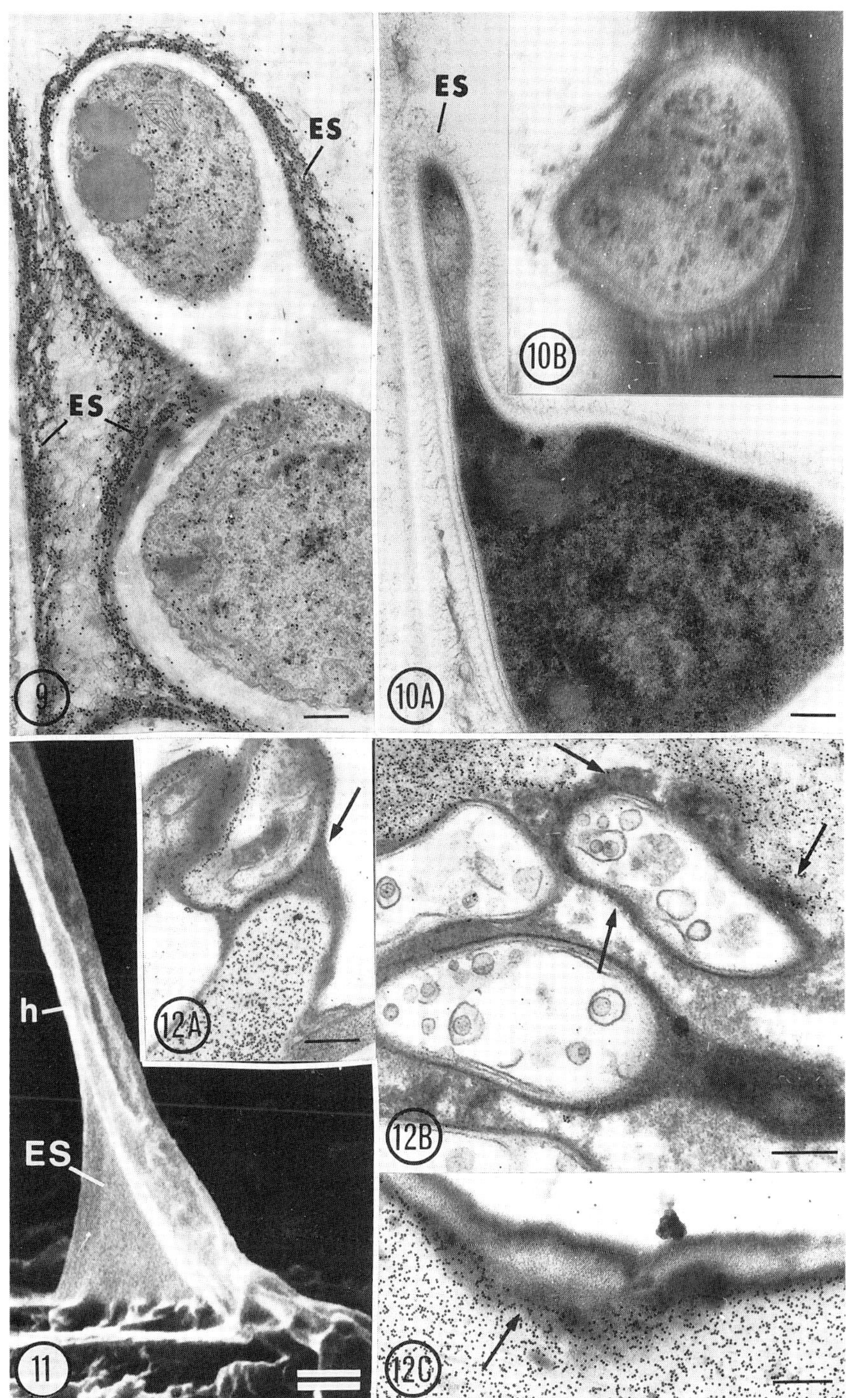
ES
ES
ES
9
ES
10B
10A
h
ES
11
12A
12B
12C

using time-lapse cinematography and material fixed sequentially after inoculation for TEM observations, described the most salient features of MH. Approaching 0.2 μm in diameter, developing MH were devoid, at least in part, of a detectable wall (Fig. 3), and contained a few ribosomes, odd strands of endoplasmic reticulum, occasional microtubules, and membranous networks (157). Host wall penetration by MH is accompanied by the formation of a halo of electron opaque material that chisels through the wall some distance from the microhypha. This halo constitutes the ensuing margin of the cavity (Fig. 3). Fibrillar strands, membranous, and tubular-like structures often extend from the fungal cell into the host wall or in the formed cavity. Layers of extracellular membranes have also been observed surrounding cells of basidiomycetes associated with degradation of wood cell walls (169).

TEM observations of *O. ulmi* development in living elm branches or in sterilized wood sections and of *P. noxius* in wood sections (Figs. 4, 5, 6) have shown that the MH of these fungi have essentially the same characteristics as those of soft rot fungi (201). The use of colloidal gold-labeled lectins, enzymes, and antibodies has led to a better knowledge of the fine structure of MH and of their association with host wall alterations. For example, probes for host wall and fungal constituents, including the plasmalemma and wall layers (47, 360), confirmed the absence of a delimiting fungal wall layer (Figs. 4, 5).

Formation of MH is mainly known in woody plant hosts, and predominantly in secondary cell wall layers of secondary xylem. Recently, however, Ouellette and Baayen (unpublished) have studied the formation of MH by *Fusarium oxysporum* f. sp. *dianthi* developing in a susceptible carnation cultivar. This wilt-causing organism breeches and spreads through and within cell walls of vessel elements (primary and secondary xylem) and adjacent parenchyma cells, by means of long MH (Figs. 7, 8) that are also very small in diameter and often devoid of a detectable wall. A very thin wall, however, can be present (Fig. 7). Profuse wall colonization (Fig. 8) appears to be directly related to symptom expression resulting from extensive tissue disruption; living cells are not appreciably affected until their walls rupture.

EXTRACELLULAR SHEATHS

Extracellular sheaths (ES) associated with fungal cells were detected as early as 1963 (291). Recently, numerous investigations have been devoted to ES produced by various fungi, including pathogenic (113, 346), saprophytic (421), endophytic (506), and mycorrhyzal fungi (183). Electron microscopy and recent cytochemical techniques have provided direct information on the roles ES may play in host-fungi interactions.

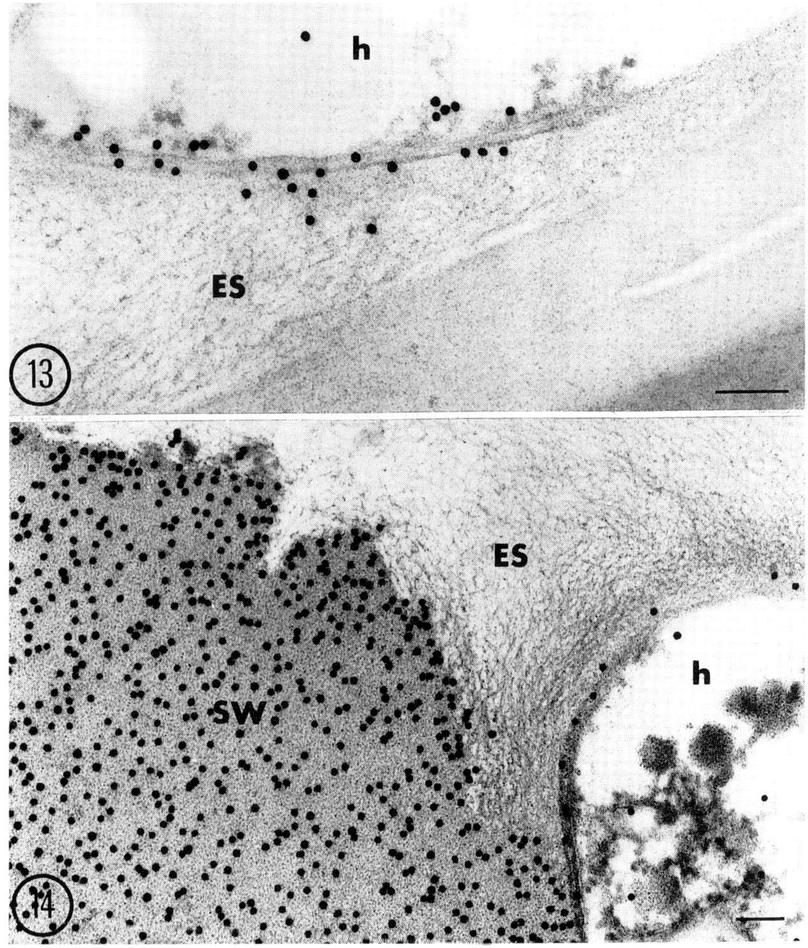

Figs. 13–14. TEM observations of *Rigidoporus lignosus* inoculated onto sterilized birch (*Betula papyrifera*) sections. **13**. Immunolocalization of laccase L1 over hyphal (h) cell wall and adjacent ES fibrils. Bar: 0.1 μm. **14**. Close association of ES fibrils with host secondary wall (SW) degradation. SW labeled for cellulose. Bar: 0.1 μm.

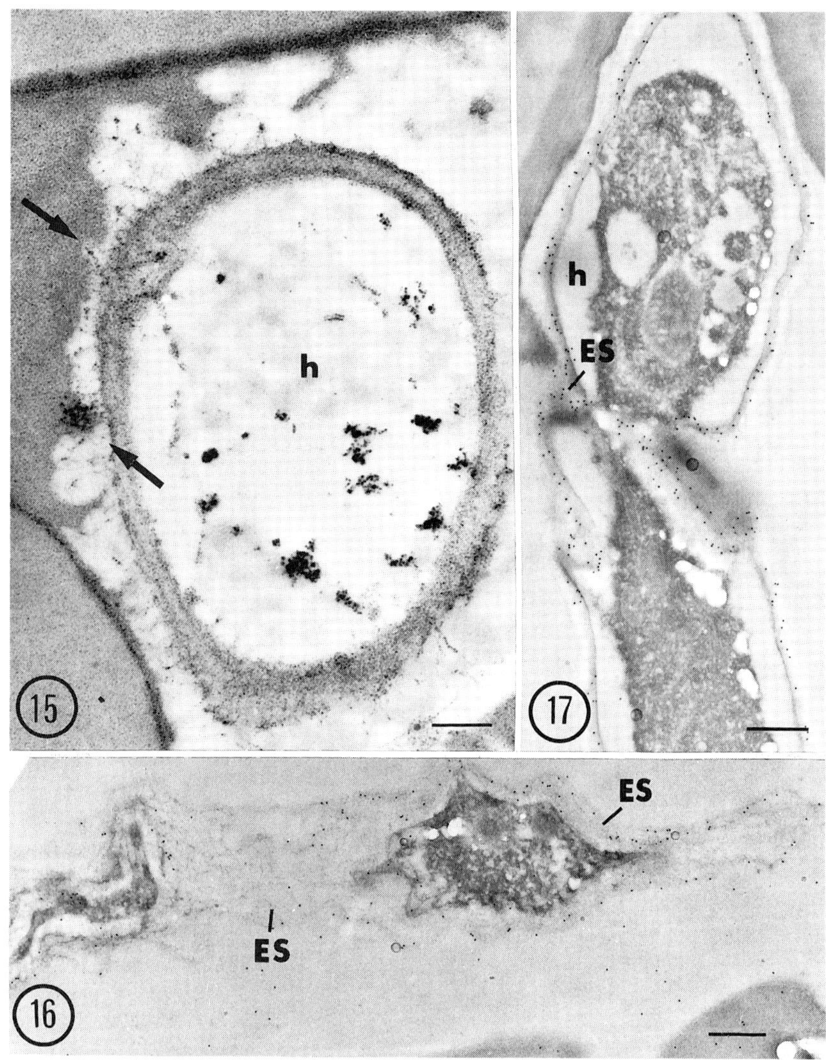

Figs. 15–17. 15. *Phanerochaete chrysosporium* in birch wood; P.A.T.Ag staining. ES of the hypha (h) is in close association with degraded parts of HCW (arrows). Scale bar: 0.2 µm. **16.** *Phlebia radiata* in birch wood. Immunocytochemical labeling with anti-lignin peroxidase (LiP). The hyphal cells shown are surrounded (and bridged) by ES, which shows a strong positive reaction for LiP. Fixed with 2% glutar-and paraformaldehyde. Scale bar: 0.3 µm. **17.** *Phanerochaete chrysosporium* degrading *Populus* wood. Immunolabeling with anti-xylanase. A positive reaction for xylanases is evident over the wall of the hypha (h) and portions of its ES located in close contact with the degraded HCW. Scale bar: 0.4 µm.

Occurrence of ES in fungi

A wide range of fungi produce a conspicuous extracellular layer of material. ES occur within many groups of fungi, from protoctista, i.e. *Peronospora* and *Sclerospora*, to higher groups, including obligate parasites (*Cronartium* and *Puccinia*), wood-rotting fungi (*Phanerochaete*, *Rigidoporus*), saprobes, and mycorrhyzal fungi. Generally, ES formation seem to occur during a certain period of the fungal life cycle or in response to particular physiological conditions, independently of the type of disease they cause.

Ultrastructure

Extracellular sheaths coat several fungal structures such as appressoria, haustoria, conidia, and penetrating hyphae (117, 162, 327), thus displaying variations in structure and morphology. In most cases, ES appear as a fibrillar network, with dense or loosened fibrils arranged around the hyphal cell, either located at tips or along the length of hyphae (Figs. 9-12C). In most cases, ES adhere to the fungal structure and can establish a close junction between the mycelium and HCW (363, 421) (Figs. 9, 13-14).

Chemical Constitution

Chemical and Biochemical data

Most information concerning the chemical nature of ES has been gained from studies on lignolytic Basidiomycetes. These fungi are able to produce large ES when grown in starvation conditions (471). Formation of ES is dependent on growth conditions that influence the secondary metabolism of the fungus (51, 421).

Fungal ES have been shown to contain a β-1,3-linked glucan main chain, more or less substituted with single glucose units or short glucosyl side chains attached by β-1,6 linkages (51, 87, 144, 421). Slight structural differences between the cell wall and the ES glucans have been demonstrated by β-1,3 glucanase degradation. The enzymatic digestion produced different ratios of glucose and gentiobiose according to the substitution degree of the β-1,3-linked backbone with β-1,6-glucosyl residues. The adherent glucan was less substituted by 1,6 glucose units than the free glucan in the culture medium (388). Similar analyses have indicated that the ES produced by *P. chrysosporium* and liberated into the medium (87) has a highly branched structure.

Surface proteinaceous fibrils, called "fimbriae" because of their similarity with bacterial fimbriae, have also been described in several fungi (130, 393). These fimbriae consist of glycoproteins characteristic of yeast-like fungi (179). Other non-enzymatic proteins were found extracellularly (156), some of them containing a proline-rich fraction (341). Enzymatic proteins were also associated with ES of *Sphacelia sorghi* (136). Binding of proteins to the glucan was non-covalent. Recently, immunocytochemical techniques have confirmed these findings and allowed identification of several ES-associated enzymes.

Cytochemical Data

(Poly)saccharides, enzymes, and non enzymatic proteins have all been detected in ES. Using the P.A.T.Ag. stain, β-1,6 glycosyl residues were observed in several Basidiomycetes (211, 346, 421) (Fig. 15). Immunogold (421) and fluorescence labeling (88) have been used to detect β-1,3-glucans within ES of Ascomycetes and Basidiomycetes. This compound was also detected in ES of *Ophiostoma ulmi* (Chamberland, this volume). Chitin was never localized in ES of rust fungi (106, 113) or of rotting fungi (343), although it is abundant in their cell walls.

Numerous oligosaccharides have been identified by means of fluorescent- or gold-lectin probes. ConA-binding sites, indicating

---

Figs. 18–21. HCW pervasion and degradation associated with bands and masses of electron opaque material (OM)(arrows). **18A, 18B.** Metaxylem vessel elements invaded by *Ophiostoma ulmi*, at the tip of developing streaks. Scale bar: 7 µm. **19A.** Corresponding view in TEM of OM chiseling through and eroding the wall of vessel element. Scale bar: 0.4 µm. **19B.** High magnification of OM associated with HCW (middle lamella and secondary layer) alteration. Note integrity of HCW where OM is discontinuous (small arrows) and composition of the latter, opaque bodies associated with a homogenous fibrillar matrix. Scale bar: 0.1 µm. **20.** Margins of several cavities in proto- and metaxylem cell walls of susceptible carnation infected by *Fusarium oxysporum* f. sp. *dianthi*, demarcated by bands of OM (small arrows). A typical fungal cell (F), with two distinct wall layers, abutting on OM in portions close to HCW. Another cavity (B) contains what appears to be an apparent fungal cell portion (arrow head). HCW labeled for cellulose. Scale bar: 0.4 µm. **21.** Strands of opaque material (small arrows) extending into HCW (vessel element) from a layer of similar material coating the vessel wall. On the opposite side of the vessel lumen a fungal cell present in continuity with a small or collapsed hypha that was similar in appearance to the structure shown at F. A microhypha is present (arrow head) in the vessel wall. Scale bar: 0.3 µm.

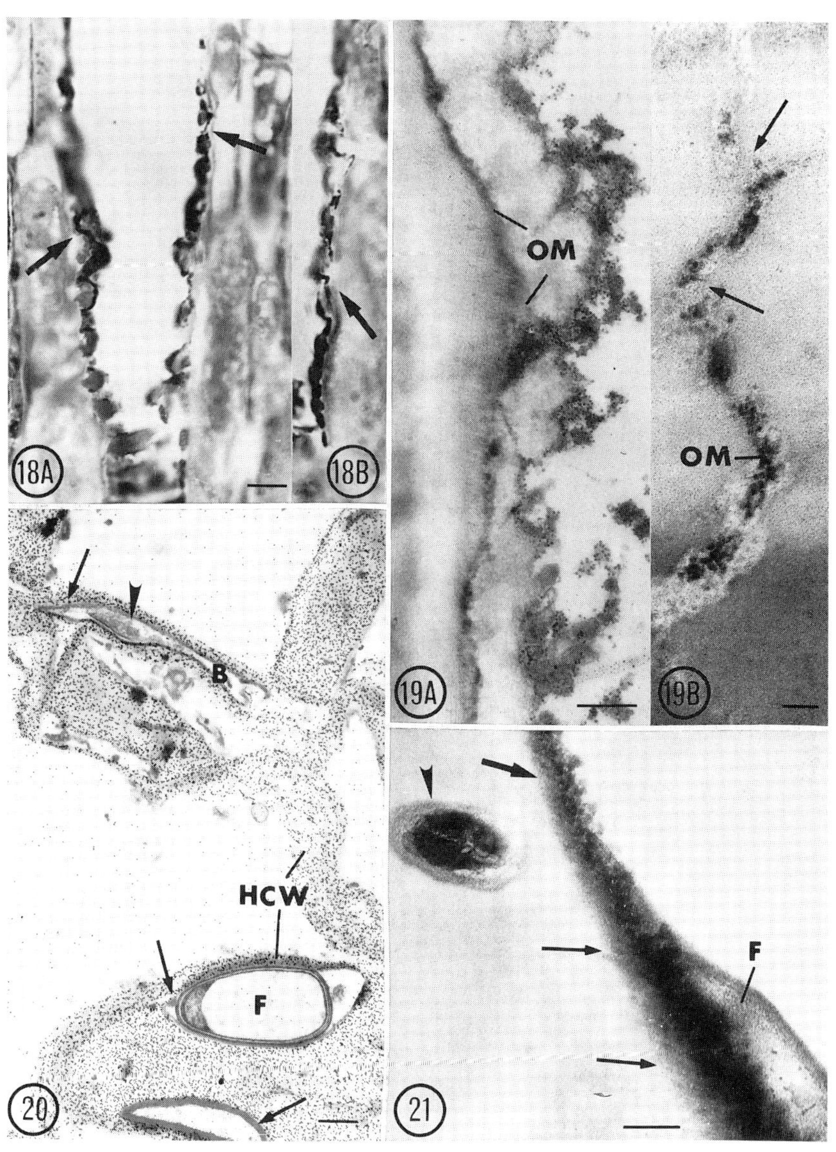
18A
18B
OM
OM
19A
19B
B
HCW
F
20
F
21

the presence of α-manno- or glucopyranosyl residues appear to be common within ES of rust fungi (113), and fungal symbionts (75). Other sugar residues such as galactose (306, 346), N-acetyl-D-galactosamine (306) and sialic acid (46) were also detected in various fungal ES. These sugar residues are probably present in complex forms as glycoproteins, thought to be involved in cell-cell adhesion and recognition. Antibodies raised against fimbrial proteins of *Ustilago* sp. were used to characterize such molecules in ES (48, 403). Localization of RNA in ES of *Ascocalyx abietina* (=*Gremmeniella abietina*) indicates that cytoplasmic constituents from fungal origin may be present in the sheath (46).

A strong acid phosphatase activity was observed in ES of endomycorrhyzal fungi during HCW penetration (183). Three major types of wall-degrading proteins were localized in ES wood-rotting fungi: (i) peroxidases, such as lignin and manganese peroxidases (65, 127, 421) (Fig. 16), (ii) laccase (344) (Fig. 13), and (iii) glycohydrolases, such as glucosidase (175), endo- and exoglucanases (421) and β-1,4-xylanase (193, 422) (Fig. 17). Microscopic observations revealed that these enzymes can be located over fibrils close to the fungal cell wall or over fibrils that were seen inside HCW (343, 346, 421). In addition, P.A.T.Ag staining performed after gold colloidal labeling strongly suggests a close association of lignin peroxidases with the glucan fibrils (421).

Figs. 22–25. 22. A *Fusarium oxysporum* f. sp. *vasinfectum* cell in close contact with a wall of a cotton xylem cell labeled for cellulose. Portions of fungal cell wall (hc) seem to penetrate HCW (arrows) by means of tubular-like structures, some of which appear to extend from within the fungal cell (arrow heads). Scale bar: 0.1 µm. **23.** *Ophiostoma ulmi* in sterilized *Ulmus campestris* wood sections. Fixation by high pressure freezing. Several small to large cavities (arrows) in HCW associated with OM, the latter encompassing portions of HCW (small arrows). Presence of gold particles indicate positive labeling for cellulose. Scale bar: 0.4µm. **24.** Bands of OM, extending intercellularly, at one point through the membrane of a half-bordered pit (arrow); OM bifurcating towards a secondary wall and possibly in continuity at one point with the smaller OM band present in this wall layer (arrowhead). This band extended between several more cells. Artificially inoculated American elm. Scale bar: 0.5 µm. **25.** A microhypha of *O. ulmi* growing in sterilized *U. campestris* wood sections, fixed with high pressure freezing. This hypha contains solely fibrillar material and opaque bodies, the size of those observed in OM described above. Scale bar: 0.3 µm.

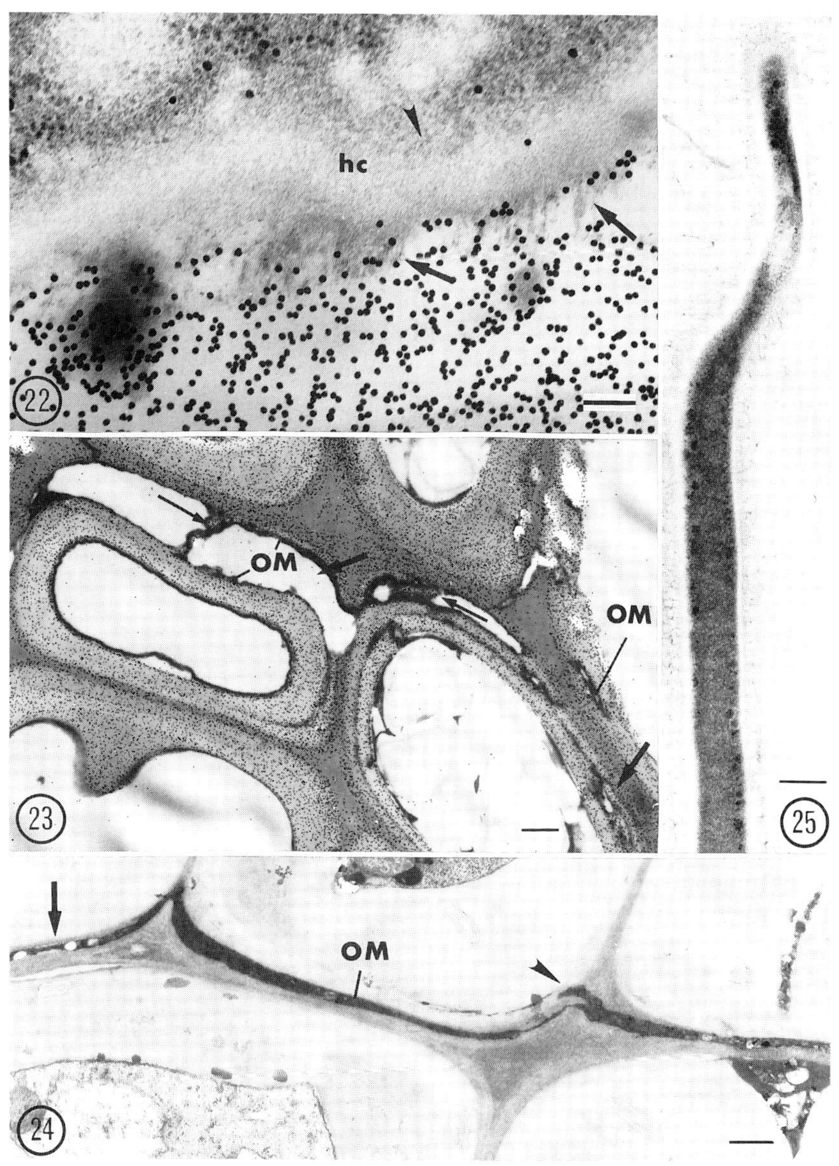
hc
22
OM
OM
OM
23
25
OM
24

Origin of extracellular sheaths

The origin of fungal ES is poorly documented as compared with their chemical constitution. Bullock et al. (88) suggested that ES of *Sclerotinia minor* result from secretion of fibrillar material through the fungal wall. In contrast, work on ES of wood-rotting fungi (346, 421) revealed close connections between ES and external portions of the fungal cell wall, suggesting that ES could originate from fungal wall components. This observation is consistent with the fact that β-1,3-glucans are constitutive components of both the fungal cell wall and ES (421). Likewise, it has been suggested that extracellular mycofibrils of *Postia placenta* originate from the hyphal surface (285). The extracellular matrix of some rust fungi also appears connected to the haustorial wall but not to the extrahaustorial membrane. In *F. oxysporum* f. sp. *dianthi*, ES is formed as an apparent extension from the outside cell wall layer, with indications that it may also be linked with cell contents (Figs. 12A–C). This ES can also pervade HCW. Proteinaceous fimbriae are thought to be assembled in the fungal cytoplasm and extruded through the wall (393). This was confirmed by immunocytochemical observations (47, 179). On the other hand, Gardiner and Day (179) showed that fimbriae could appear either as extruded long fibrils or as a short fringe embedded in the wall.

Roles of extracellular sheaths

The discovery of extracellular proteins and glycoproteins within rust ES (113, 156, 228, 322) has suggested that they may be involved in binding fungal structures to plant surfaces. Sugars, such as glucose, mannose, fucose, and galactose, characterized in ES and cell wall of several fungi (46, 113, 306, 346) may also play an important role in specific attachment in plant-fungi systems, being bound by plant surface lectins.

Microscopic observations of fungal ES when hyphae were localized in the vicinity of HCW revealed that the ES network plays an essential role in wall degradation. During colonization of living trees or inert wood (336, 343, 421), ES preceded wall penetration by hyphae. Attack of xylem cells was associated with fungal fibrils that extended from hyphae into HCW (Fig. 14) (346). In advanced decay, sheath fibrils were closely associated with detached portions of HCWs, even surrounding and penetrating them (336, 343, 421). At times, ES alone was attached to degraded HCWs (421).

Lignin peroxidases (65, 127, 422, 466), xylanase (193), laccase

(344), and β-glucosidase (175) have been localized within ES of various rotting fungi, as have esterases, cellulases, pectinases, invertase (317), and cutinase (265) in *Colletotrichum* sp. The non covalent loose association between ES and enzymes (193, 421) may support a translocation by ES. The presence of associated β-1,3-glucanases and oxidases (122, 128, 246, 344) that can generate hydrogen peroxide necessary for the function of lignin and manganese peroxidases indicates that ES can be considered as a part of the lignin-degrading system of fungi. Uneven cytodetection of such enzymes in fungal ES, however, suggests their localization in ES may depend upon (i) the enzyme substrate accessibility, (ii) the pathogen environment, and (iii) the physiological condition of fungi.

Several additional roles have been suggested for fungal ES. Oligosaccharides of fungal glucans possibly linked with ES are known to be potent elicitors of phytoalexins (11, 107, 441). In wood-rotting fungi, polysaccharides and proteinaceous compounds of ES may provide nutrition for growing hyphae (471). Fungal ES may also protect fungal cells from unfavorable environmental conditions. For example, the fungal laccase found within *Rigidoporus lignosus* ES (344) polymerizes fungitoxic phenols released from lignin degradation (182).

OTHER SITUATIONS OF HCW ATTACK

Extensive host wall invasion and degradation have also been asssociated with masses and bands of electron opaque material (OM) (361, 362). Pronounced alterations of vessel walls, particularly of the primary and early secondary xylem, are noticeable in infection by *Ophiostoma ulmi* in elm as in nonhost trees (see Fig. 8 in Rioux and Biggs, this volume). Such alterations can be linked to the presence of variously configured OM (Figs. 18A, 18B). Ultrastructurally, such material is constituted of a fine fibrillar matrix and small particles the size of ribosomes (Figs. 19A, 19B). Cell wall breakdown and cavity formation in susceptible carnation can also be linked with bands of OM that seem analogous to the outside wall layer of the *Fusarium* pathogen cells (Fig. 20). Arrays of parallel tubular-like structures that penetrate directly into the host wall extend from these OM bands and masses.

In elm, xylem tissues adjacent to vessel elements of the secondary xylem may also be penetrated by bands of OM of various thickness and configurations with fine to larger strands projecting at angles from the main band (Figs. 23, 24, 25). These bands may extend between or within wall layers, with some portions often ending within cells (360, 361). Murmanis and colleagues (336) also hypothesize that much of the OM material originates from the fungus.

CONCLUSION

Pathogenic fungi utilise various strategies to penetrate and degrade host cell walls. Appressoria and haustoria are structures involved both in wall adhesion and penetration while fungal sheaths have been shown to play an active role in cell-cell recognition and HCW degradation. However, recent cytological studies have revealed that uncommon fungal components, such as microphyphae, tubular-like structures, or opaque material, may also contribute to HCW alteration. This constitutes a new and exciting research avenue in plant pathology.

CELL WALL CHANGES IN HOST AND NONHOST SYSTEMS: MICROSCOPIC ASPECTS

Danny RIOUX[1] and Alan R. BIGGS[2]

[1]Natural Resources Canada, Forest Service, Quebec Region, 1055 rue du P.E.P.S., Sainte-Foy, Quebec, Canada G1V 4C7
[2]West Virginia University, University Experiment Farm, P. O. Box 609, Kearneysville WV 25430

When plants are challenged by pathogens that penetrate cell walls directly, through natural openings, or wounds, the plant defense system carries out numerous metabolic modifications, some of which induce cell wall changes visible in light microscopy. These changes may appear in a few isolated cells or be present in many contiguous cells, thus forming a specialized defensive tissue.

A better understanding of host-pathogen interactions has been obtained in the last two decades by comparing host and nonhost reactions to pathogens (217, 218, 406). Here we present a brief survey of cell wall changes in herbaceous and woody hosts and nonhosts and we discuss the effect these changes may have on the colonization rate of fungal intruders.

NONHOST CONCEPT

A nonhost species is exempt of a particular disease mainly because of resistance factors, but also because of disease escape (324). Thus, each plant species is a nonhost for most known pathogens. Nonhost resistance is mainly based on nonspecific active defense mechanisms that are under multigenic control, and hence shares many similarities with horizontal resistance. Consequently, the pathogen must inhibit or avoid triggering resistance mechanisms in its host or tolerate the effects of the defense reactions (220).

It is possible in some cases to compare cell wall changes that occur in wounded plants with those seen in infected nonhosts. Similarities between these two types of nonspecific defense reactions are more obvious in woody plants. Before discussing such active defense mechanisms, the following section will deal with constitutive characteristics of the nonhosts that appear to be implicated in their resistance to infection.

PRE-INFECTION STRUCTURAL FACTORS

The relationship between resistance and constitutive chemical nature of nonhost cell walls has been little studied. The type of monomers composing lignin appears to affect the ligninolytic activity of white-rot fungi (62) and may similarly contribute to resistance to some pathogens (405); antifungal components may be released from cutin or suberin when these are degraded (244, 267), but evidence is lacking that either lignin or cutin monomers play a significant role in nonhost resistance. Pectic enzymes are probably involved in the process of colonization of susceptible plants. However, no correlation was found between susceptibility or resistance and the quantity and activity of pectinases produced by six pathogens when these were grown on host or nonhost cell walls (121).

Passive defense mechanisms of nonhosts such as the nature and quantity of nutrients available may influence the pathogen's vigor and its ability to penetrate the walls. This possibility, however, has been rarely considered to be an essential factor for successful pathogen penetration of cell walls. For instance, wilt pathogens grow in the xylem sap of nonhosts as well as in that of their hosts (254). In Dutch elm disease, the size of vessels and their distribution within growth rings may be related to susceptibility of elm cultivars to the disease (152, 316, 457). In this case, clusters of large vessel elements would facilitate a more rapid rate of colonization by the pathogen.

The features of the leaf surface have often been considered to be crucial for the success or failure of pathogens to develop specialized structures such as germ tubes, appressoria, or infection pegs. Wynn and Staples (525) have demonstrated that nonhost or cultivar resistance was attributable to the weak adherence of fungal propagules to the leaf surface or to the pathogen's failure to recognize morphological characters that permit it usually to localize sites of entry. Formation of penetration structures has been found in a few cases to be stimulated by chemicals such as nutrients but contact stimuli elicited by ridges on the cuticle or arrangement of epidermal walls have been considered to be crucial for the first steps of infection by leaf pathogens. The use of artificial surfaces has given conclusive evidence that recognition of contact stimuli may be a prerequisite for many pathogens to gain access to internal plant tissues (525).

INDUCED DEFENSE STRUCTURAL MECHANISMS

Herbaceous plants

For leaf-infecting fungi that succeed in breaching pre-infectional structural defenses of resistant cultivars or nonhosts, progression may be limited by cell wall changes that frequently occur close to pathogen cells. For instance, when *Phaseolus vulgaris* is confronted with a non-pathogen rust (*Uromyces vignae*), the walls are modified by silica deposits (215, 219) and accumulation of phenolic compounds (479) which appear to prevent haustorial formation. Silica deposits may be detected by an increased refractivity of the wall in light microscopy and also by energy dispersive X-ray microanalysis. Ryerson and Heath (426) report that changes of refractivity can also be caused by the presence of other unidentified compounds. Such modifications do not occur or are difficult to detect in compatible host-pathogen interactions between *P. vulgaris* and *U. appendiculatus* (220). Inoculations of the host with a compatible pathogen induce susceptibility of nonhosts to other non-pathogenic rust fungi (216), thus suggesting inhibition of such defense reactions by the pathogen.

Papilla formation has often been observed in leaves of nonhosts as a reaction to attempted penetration by other fungi (208, 308, 407, 445, 446). Papillae are formed by one or more substances such as callose, lignin, suberin, silicon and cellulose (6).

Nonlignified papillae may contain other phenolic compounds that are easily detected by autofluorescence. The presence of phenolic compounds in papillae was demonstrated for nonhost reactions of *Vicia faba* leaves against *Botrytis cinerea* (308) (Figs. 1 and 2). In susceptible situations with *B. fabae*, such material was not observed. Mansfield and Richardson (309), using transmission electron microscopy (TEM), could observe swelling and degradation of epidermal and mesophyll cell walls (Fig. 3), this often preceding the advance of pathogen cells. These authors suggested that pectic enzymes are probably responsible for wall degradation, although other enzymes may also be involved.

In leaves, lignification of epidermal walls adjacent to papillae is also frequently observed (407, 445). A disk-like lignified structure is formed around the papilla, and contiguous anticlinal walls also become lignified. Such lignification confers wall resistance to degradation by a commercial cellulase while delignification treatments render the walls accessible to enzyme attack as in the case of unaltered epidermal cell walls (407). Nonhost resistance of five

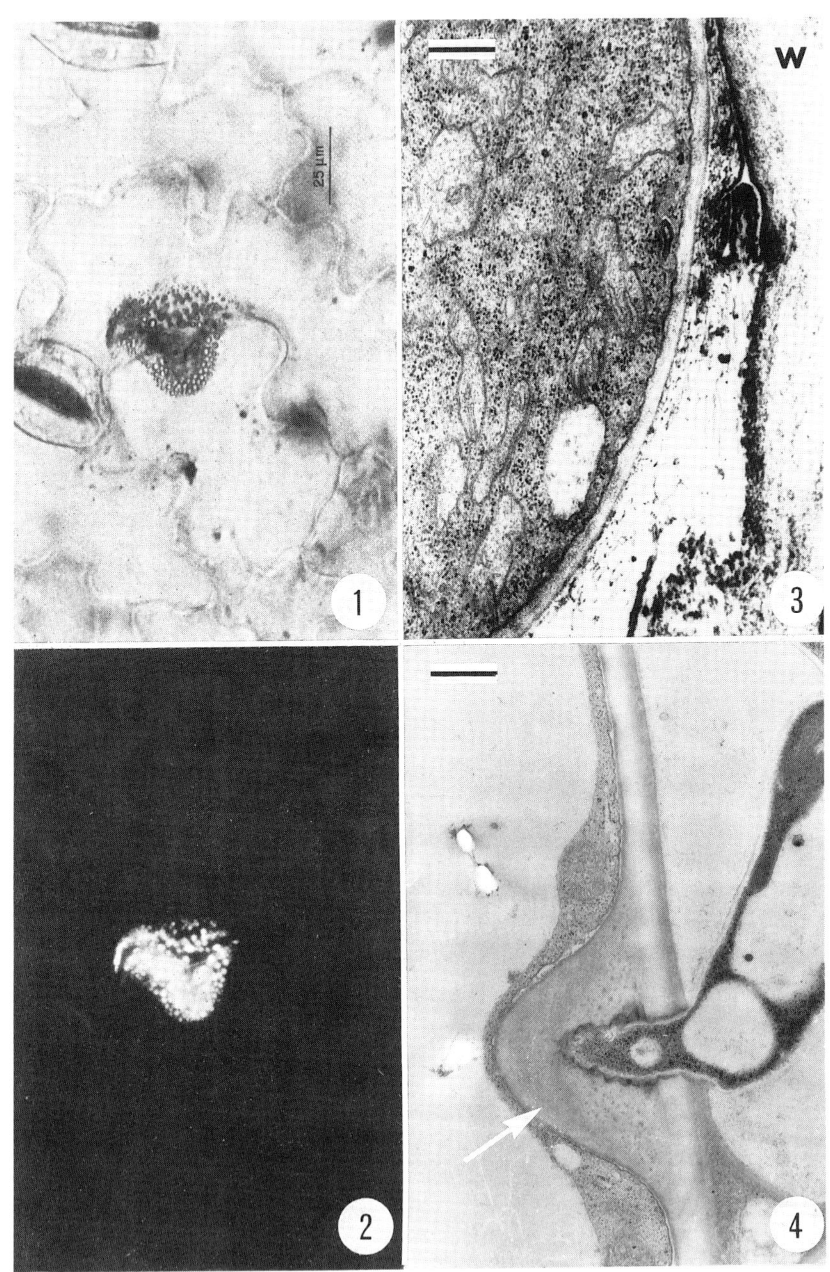
25 µm
W
1
2
3
4

species of Cucurbitaceae involves lignification of epidermal cell wall around appressoria of non-pathogens, but a similar response was not detected in compatible interactions involving *Cucumis sativus*, *Cucumis melo* and *Citrullus vulgaris* inoculated with *Colletotrichum lagenarium* (208). Results associating lignification to nonhost resistance have been obtained in studies on the reactions of potato tuber to a pathogen and two non-pathogens (207). In addition, it has been occasionally demonstrated that the induced lignin differs chemically from the lignin present in normal tissues (405). Such changes might confer greater mechanical strength to the walls or make them resistant to enzymatic degradation.

Baayen (17) inoculated stems of resistant cultivars of *Dianthus caryophyllus* with *Fusarium oxysporum* f. sp. *dianthi* (pathogen) and *F. oxysporum* f. sp. *lycopersici* (non-pathogen) and observed that the defense reactions against both pathogens were similar. In infections carried out with the non-pathogen, the invaded xylem tissue was limited to 1 cm around the inoculation point, as compared to a few cm with the pathogen. Additional wall layers and appositions were described as defensive cell wall changes, and they were frequently impregnated with phenols similar to lignin monomers. In addition, in a zone of hyperplastic parenchyma tissue, many cells differentiated and became suberized and thus formed a band apt to impede further colonization. Similar defense responses have been also observed in roots (19). In a susceptible cultivar, most of these reactions do not occur and the pathogen colonizes the stem extensively apparently following disruption of xylem cell walls (16). In herbaceous plant species, the localization of the invaded xylem (18) may resemble the compartmentalization processes often described in trees (see below). In resistant cultivars, papillae may also be involved in the defense system of leaves (445) and roots (105) (Fig. 4).

Figs. 1–4. 1, 2. A papilla in a *V. faba* leaf at a penetration site by *B. cinerea*. The autofluorescence of the granular deposit under u.v. illumination (Fig. 2) and the use of clearing solutions revealed that the papilla was nonlignified but contained other phenolic compounds. (Reprinted from 308). **3**. Degradation of a *V. faba* mesophyll cell wall (W) adjacent to an intercellular hypha of *B. fabae*. Bar = 0.4 μm. (Reprinted from 309). **4**. Papilla formation (arrow) in a tomato root cortical cell as a response to attempted penetration by *F. oxysporum* f. sp. *radicis-lycopersici*. Bar = 0.6 μm. (Reprinted from 105).

Woody plants

Studies on nonhost defense responses by woody plants are nearly absent in the literature. It is nevertheless possible to draw analogies between nonhost structural changes to infection and tree reactions to wounding as these responses often are related to the compartmentalization concept described below. This mechanism implies that many cells are formed or react to circumscribe the injured tissues. For the xylem, a model of compartmentalization was proposed to explain how the trees resist decay-causing fungi (449). This model consists of three reaction zones made of cells extant at the time of infection, which act to limit the longitudinal (wall 1), radial (wall 2) and tangential (wall 3) spread of the invader, and of barrier zone formation (BZF) that is laid down by the cambium in response to infection. The compartmentalization concept was later extended to the xylem and bark (56, 57, 63, 448). For cell wall changes that occur in infected tree leaves, which are virtually similar to those described previously for herbaceous plant species, see Adaskaveg (2).

In the xylem of the nonhost trees *Prunus pensylvanica* and *Populus balsamifera* inoculated with *Ophiostoma ulmi*, continuous barrier zones consisting of cells with additional wall layers and containing phenolic compounds were formed to enclose the invaded xylem (Fig. 5) whereas BZF was absent or discontinuous in the host *Ulmus americana* (409, 410). In addition, at 2.5 cm upward from the inoculation point, BZF was usually observed within 15 days after inoculation in the nonhosts, but not until approximately 30 days in the susceptible elm. Lignin was detected histochemically in the walls of barrier zone cells in greater concentration than elsewhere in the normal xylem tissues. Suberin was also detected with the method described by Biggs (53) and was also reported in barrier zones of other tree species (367, 368).

The vessel elements included in or contiguous to barrier zones in nonhosts often contained internal suberized wall appositions. In *P. balsamifera*, this was constantly observed and was caused by the presence of suberized tyloses. Such tyloses in this nonhost are also an important component of the reaction zone (410) that corresponds to the wall 3 region of the compartmentalization of decay in trees (CODIT) model (449). In mature tyloses, the use of an exoglucanase conjugated to colloidal gold demonstrated that the tylosis wall was multilayered (Fig. 6). Occasionally in *P. pensylvanica*, the reactions for suberin were apparently caused by a suberized, lamellar coating (Fig. 7). Such suberized structures probably provide the walls with a greater resistance to microbial attacks, although some decay fungi can penetrate tylosis walls, the composition of which, however, has not been described (63).

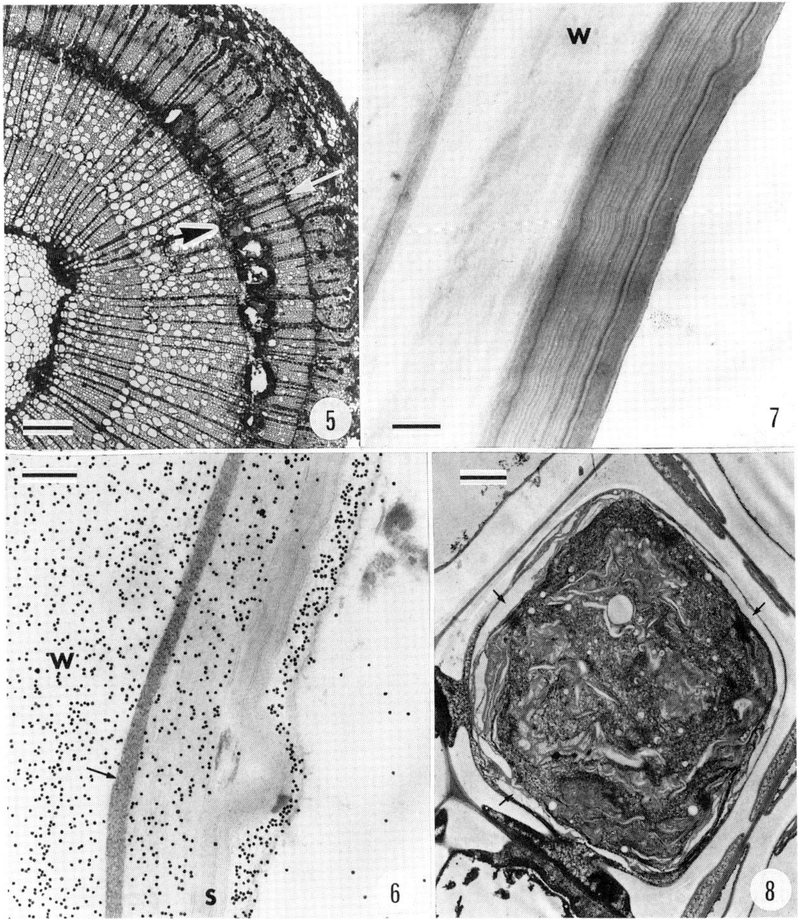

Figs. 5–8. 5. A continuous barrier zone (black and white arrow) in *P. pensylvanica* surrounding xylem invaded by *O. ulmi*. Cambium (white arrow). Bar = 200 μm. **6**. Adjacent to a vessel secondary wall (W) of *P. balsamifera* is a multilayered tylosis wall constituted of an external layer (arrow) and a suberized layer (S) giving negative labeling with an exoglucanase conjugated to colloidal gold; this layer is comprised between two layers positively labeled for cellulose. Bar = 0.2 μm. **7**. Secondary wall (W) of a vessel of *P. pensylvanica* coated with a suberized layer with an obvious lamellar structure. Bar = 0.3 μm. **8**. Wall fragments (arrows) chiselled through by bands of opaque material are evident in this occluded vessel of *P. pensylvanica*. Bar = 1.5 μm.

Formation of tyloses has been reported in elm affected by Dutch elm disease but their development seems to occur too late after fungal invasion to play a significant role in resistance (361). Tyloses were also reported in elm inoculated with a non-pathogen (152, 153). Suberized tyloses were observed in elm following inoculation with *O. ulmi*. In such cases, however, they were not localized in strategic areas such as the barrier zone or the wall 3 reaction zone, as is usually the case in *P. balsamifera* (410). Suberized tyloses were observed as a response to wounding in many tree species (55).

Rioux and Ouellette (409) compared 13 nonhost species inoculated with *O. ulmi*, considering factors such as the extent of colonization, the intensity of histological changes and attempts to reisolate the pathogen, and reported *P. pensylvanica* to be the most susceptible nonhost to *O. ulmi*. In this nonhost, alterations of vessel walls were seen regularly, occasionally after direct penetration of the fungus but usually in the presence of various substances that could occlude completely vessel lumina (Fig. 8). The xylem cell wall alterations commonly observed in infected susceptible elms are often associated with an electron-opaque compound (361). In the nonhost *P. balsamifera*, such alterations were unusual and papillae, in addition to tylosis formation, were formed at times in parenchyma cells adjacent to vessel elements containing fungal cells (Fig. 9). At times, some papillae were formed in *U. americana* in response to penetration by *O. ulmi* (Fig. 10). Numerous papillae were observed when this elm was injected with fungicides (Ouellette, unpublished results). Thus, fungicides might act either directly or by stimulating host defense reactions.

In *P. balsamifera*, some pith and perimedullary zone cells dedifferentiated and became suberized following inoculations with *O. ulmi* (Figs. 11 and 12). Since suberin has been detected in such barriers (410) and since pathogens like *Cronartium ribicola* and *Leucostoma persoonii* may be observed in the pith of a resistant pine (243) and of susceptible *Prunus persica* (54), respectively, it is possible that the role of such medullary tissues has been overlooked in the past. These tissues may play an important role in limiting or in favoring pathogen colonization.

In nonhost controls injected with sterile distilled water, most of the nonspecific responses described above were also seen, but they were formed just within 2-3 mm around the inoculation wound and they were less numerous and more restricted than in infected nonhosts (409, 410). Thus, the types of reaction observed in the xylem in the immediate vicinity of the wound, but not necessarily their intensity, may be indicative of what will happen when a woody plant is challenged by non-pathogens. In roots, extensive BZF has also

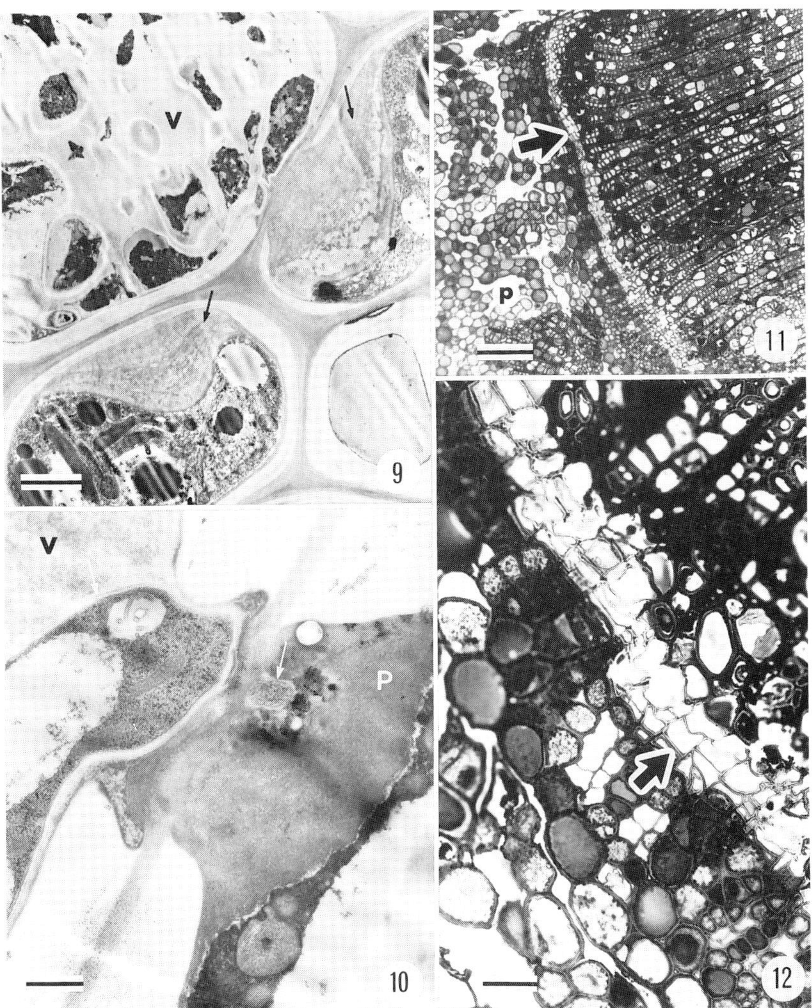

Figs. 9–12. 9. In *P. balsamifera*, papilla formation (arrows) in two ray cells adjacent to a vessel (V) containing *O. ulmi* cells. Bar = 3 μm. **10**. Papilla formation (P) in a parenchyma cell of *U. americana* in response to attempted penetration by *O. ulmi*. Fungal cell (arrows). Bar = 1 μm. (Courtesy G. B. Ouellette). **11**. A band of suberized cells (arrow) has been formed in *P. balsamifera* between the invaded xylem (at right) and the pith (P). Bar = 150 μm. **12**. At higher magnification, the aspect of such suberized cells (arrow) resembles that of phellem cells. Bar = 25 μm.

been attributed to the presence of different pathogens whereas this response was more limited in reaction to wounding alone (491, 492).

In bark, the deposition of polysaccharides, including callose, was one of the first observable histochemical reactions to wounding (56). Within 72 hr, cell wall lignification internal to the area of polysaccharide deposition occurred, followed by suberin deposition in lignified cells about 24 to 48 hr later (Fig. 13). A meristematic layer formed immediately internally and contiguously to the ligno-suberized layer (LSL) within 24 to 48 hr after its appearance (Figs. 14 and 15), and wound (necrophylactic) periderm (NP) often was formed 10 days after wounding (Fig. 16). Complete formation of the LSL and NP around the entire wound occurred within 14 to 21 days under ideal conditions.

The delimitation of fungal pathogens in bark of woody plants ultimately occurs via the formation of the LSL derived from cells extant at inoculation and NP derived from newly differentiated phellogen (56, 57). For annual cankers, these barriers are effective in preventing continued colonization by the fungus and diseased tissues are usually sloughed. In perennial cankers, bark boundary tissues, which may include extensive xylem formation, are only partially effective and are either directly penetrated or circumvented annually, resulting in a series of concentric callus ridges.

Infection of bark wounds was greatest when the inoculum arrived at the infection court immediately (78, 408, 423). In older wounds, disease frequency and severity declined with time until the wounded tissues expressed resistance comparable to that of noninjured bark. Receptiveness of the infection court to inoculum was influenced by external events such as colonization by nonpathogenic epiphytic organisms, as well as by internal events such as host boundary setting processes. Histological studies with peach bark and *Leucostoma persoonii* showed that wounds resistant to inoculation possessed a minimum of three phellem cells in the NP. At earlier stages, wounds were susceptible to the fungus, although severity of the symptoms declined beginning about 3 days after wounding. The presence of NP was critical for inhibition of the pathogen, whereas lignified cells and/or the LSL significantly slowed fungal colonization. No chemical or morphological barriers were observed 3 days after wounding, suggesting a role in resistance for phytoalexins or other substances in the early stages of the wound reaction.

The relative number of cells in suberized periderm or the thickness of the suberized layers in wounded potato tubers were generally positively correlated with resistance to pathogens. In

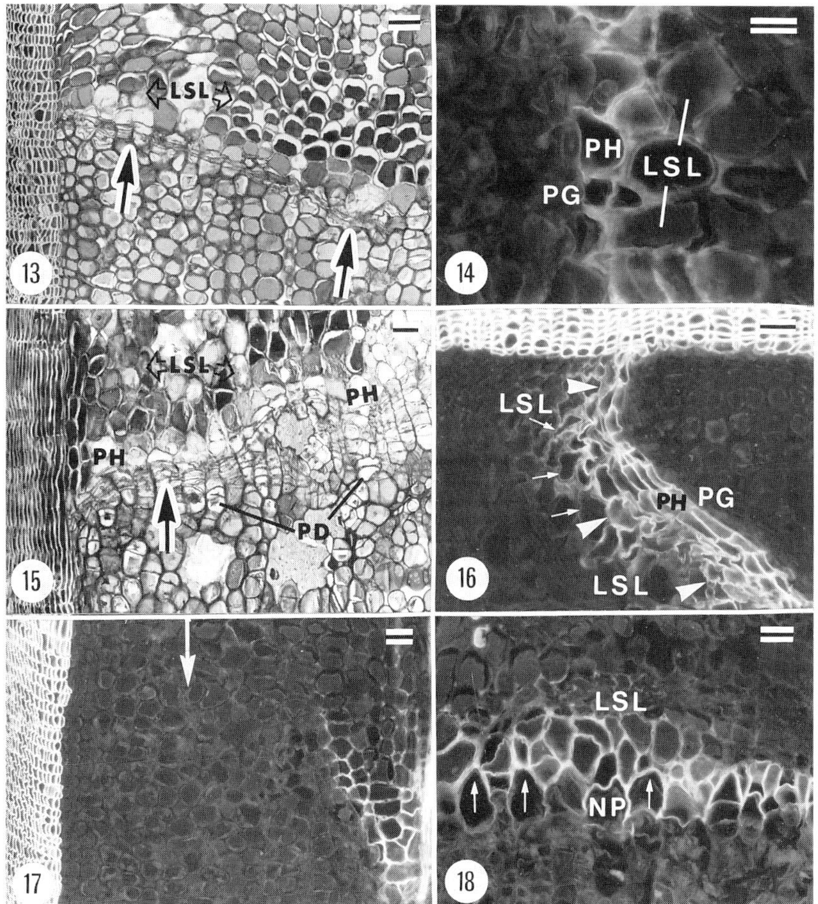

Figs. 13–18. 13. Section of differentiated phellogen (arrows) and incipient ligno-suberized layer (LSL and open arrows) 96 hr after wounding. **14**. Bark tissue stained with phosphine under ultraviolet excitation showing LSL, first phellem (PH), and new phellogen (PG), 7 days after wounding. **15**. Transverse section of differentiated PG (arrow) showing PH, phelloderm (PD), and the LSL, 7 days after wounding. **16**. Bark tissue 10 days after wounding stained with phosphine under ultraviolet excitation showing new PG, differentiated PH, and cells of the LSL (its internal boundary delimited by arrowheads) exhibiting cell wall lignification and suberin linings (arrows). **17**. Cortex region of bark in advance of colonizing hyphae (direction indicated by arrow) of *Leucostoma persoonii* showing suppression of tissue changes associated usually with formation of LSL and necrophylactic periderm (NP). **18**. Portion of LSL and NP in bark tissue colonized by *L. persoonii* (top) showing altered cell shape (arrows). Scale bars: 10 μm.

peach, resistance to *Leucostoma* spp. was associated with increased rate and amount of suberin accumulation rather than with the number of suberized cells or thickness of suberized layers. Lignin accumulation, which is thought to play an important role in many host/pathogen interactions (499), appears to be less significant than suberin in peach bark and perhaps in the bark of other tree species.

The formation of LSL and NP in bark inoculated with pathogens is usually similar to that observed in noninoculated wounds or wounds inoculated with non-pathogens, although there are a number of notable differences (57). Pathogens affect the position or location of NP and related tissues in wounded bark relative to their location in the absence of pathogens (Fig. 17). In wounds, new tissues are regenerated usually within 1-2 mm of the wound surface. When wounds are colonized by pathogens, new tissues may be formed several cm from the infection court. Pathogens affect the morphology and differentiation of new tissues and may retard or inhibit the formation of NP and/or the deposition of suberin in wound periderm and xylem barrier zones. NP that formed in peach bark inoculated with *Leucostoma* spp. exhibited phellem cells which were thin-walled and rounded, as compared with the relatively thick, rectangular phellem produced at control wounds (Fig. 18). Wounds inoculated with *Leucostoma* spp. and *Botryosphaeria* spp. showed delayed or inhibited suberin formation in cells of the ventral callus surface (57). This region was often a site for continued colonization of bark and xylem tissues by the pathogens. When xylem tissues were examined for suberin formation following wounding, large areas of suberized ray parenchyma were observed in the wall 3 region based on the CODIT model (449). Such regions were absent in inoculated xylem (54).

Few woody hosts are able to challenge successfully the initial invasion of a pathogen with the formation of a single LSL and NP, as it occurs following wounding or inoculation with nonpathogens. Instead, the tree often produces a series of incomplete periderm layers within a single season in response to the advancing fungus. The fungus will continue to spread in the host tissue until conditions become more favorable for the host to form a complete periderm and less favorable for the pathogen to penetrate further. In bark of peach cultivars inoculated with *Leucostoma* spp., NP formed 21-28 days after inoculation and was associated with a delimited lesion in both cultivars inoculated with *L. cincta*, the less pathogenic fungus. For cultivars inoculated with *L. persoonii*, the wound periderm was not as thick, had fewer cells, and was associated with only a slight inhibition of lesion expansion followed by a second rapid increase in canker length as the NP was breached 28-35 days after inoculation

(54).

Few studies have examined the responses of bark tissues inoculated with non-pathogens. In peach bark that was wounded and inoculated with *Alternaria alternata*, *Trichoderma harzianum*, and *Epicoccum nigrum* alone and in combination with the pathogen *L. cincta*, the course of events described above was observed within the normal time frame in all of the control wounds and with only one of the three non-pathogens (58). Inoculation of wounds with *E. nigrum* alone, or in combination with *L. cincta*, did not cause any deleterious effects on the wound responses examined. Inoculation of wounds with *T. harzianum* and *A. alternata* resulted in delayed LSL and NP formation. Abnormalities in LSL and NP formation associated with pathogenesis and described above were not observed following inoculations with the two colonizing non-pathogens, except for changes in the location and rate of differentiation of LSL and NP relative to the control.

The direct involvement of the LSL and/or NP, or of by-products generated during the differentiation of these tissues, in the resistance of woody plants to fungi has never been demonstrated conclusively. However, several lines of evidence indicate that these tissues prevent desiccation of internal layers and inward movement of pathogen toxins or enzymes, and serve as physical barriers to colonization. The generation of new tissues may also result in lignification of pathogen hyphae (54) and/or in the production of phytoalexins. Despite the abundance of correlative data presented in recent studies, the precise role and importance of ligno-suberized tissue and new periderm, and their biochemical components, in resistance to fungal pathogens is still poorly understood.

CONCLUDING REMARKS

Many reactions of nonhosts counteract the effects of infection or wounding. Although their occurrence has often been correlated with resistance to different diseases, chemical changes or other undetected structural changes may be the primary factors permitting nonhosts or resistant cultivars to withstand such injuries. Even the use of chemicals, such as cycloheximide, that inhibit cell wall changes (217, 446) and induce susceptibility to subsequent inoculations are not an irrefutable proof of their importance, since these chemicals may also influence other important biochemical pathways.

Modifications of cell walls by compounds such as lignin, suberin, callose, etc. have been considered to limit direct pathogen colonization either in conferring greater strength to the wall, in masking or rendering the polysaccharides unsuitable for degradation by cell wall-degrading enzymes, or in releasing fungitoxic substances upon degradation. They may also act indirectly by increasing the impermeability of the wall, which may result in a greater concentration of phytoalexins localized in the immediate vicinity of pathogen cells (218). Impermeability also may hinder the diffusion of toxins or prevent the pathogen from obtaining essential plant nutrients. On the other hand, some investigators have reported that the reactions of compartmentalization in the tree were formed mainly to maintain sapwood function (72), the conditions that occur in functional sapwood *per se* being inadequate for the growth of fungi. This situation is further complicated when one considers that each mechanism must be present at the right time, at the right place, and with sufficient intensity.

Even if some difficulties prevent a consensus on the importance of these nonspecific mechanisms, it is evident that it would be worthwhile to exploit the potential of nonhosts to improve plant disease resistance. Since these mechanisms are under polygenic control and since it is still nearly impossible to cross different species, it may seem difficult to transfer useful characters to susceptible species. A few studies, however, have demonstrated useful levels of heritability for compartmentalization in the xylem (297, 432), and suberin deposition in the bark NP (59). Molecular engineering aimed at the modification of only one or two major genes implicated in the regulation of mechanisms such as phytoalexin production and BZF might give just enough time for a plant to defend itself against the attack by a pathogen.

WOOD POLYSACCHARIDES AND THEIR DEGRADATION BY FUNGI

Jean-Paul JOSELEAU and Katia RUEL

Centre de Recherches sur les Macromolécules Végétales, CNRS, B.P. 53X, F–38041 Grenoble cedex, France

Plant cell walls possess a microfibrillar architecture based on a polysaccharide lattice. The chemical and physical nature of the wall varies according to plant group and cell type, but all plant cell walls contain several polymer components and are multilayered. The architecture of the cell wall of land plants is organized around cellulose macromolecules aggregated together in bundles of crystalline microfibrils, embedded in a matrix of non cristalline polysaccharides, the hemicelluloses, and, in higher plants, of a complex phenyl-propane polymer, lignin. In plant cell walls polysaccharides are most abundant, lignin accounting for as much as about 30% of the secondary wall of the most lignified tissues.

During the first infection and colonisation events, pathogenic and saprophytic fungi attack the host cell walls. The enzymatic array produced by a fungus determines its ability to degrade the cell wall. Different patterns of degradation of the plant cell wall may be observed, depending on the more or less selective degradation of lignin, cellulose and/or hemicelluloses (157). Correlations between the nature of predominating enzymes and the pattern of the attack by white-rot fungi has been established by electron microscopy using specific substrate staining and immunocytochemical labeling of the enzymes (418).

Here we describe the biochemical composition of the lignified cell wall, limiting this review to the polysaccharides commonly found in wood, i.e. cellulose and the hemicelluloses of the family of xylans and glucomannans. We also describe the mechanisms of degradation of such polysaccharides by fungi.

WOOD CELL WALL POLYSACCHARIDES

Cellulose

Cellulose is the basic constituent of all plant cell walls in which it represents 20 to more than 90% of the dry weight. In its native form, cellulose is largely crystalline, the glucan chains being held together by hydrogen bonds distributed along and between the chains in a regular arrangement to yield crystalline fibers. This structure is formally described as a two-fold extended helix (13). Intra and interchain hydrogen bonds maintain the crystalline structure. The dimensions of the microfibrils vary according to the origin of the cellulose (102) which can exist under different polymorphic forms after chemical or mechanical treatments that modify the crystal lattice. *In situ* microstructures of native cellulose in wood have been investigated using solid state ^{13}C NMR spectroscopy (501).

Xylans

Xylans are the most abundant polysaccharides after cellulose, accounting for up to 30% of the dry weight in hardwoods. Xylans are heteropolysaccharides that include a wide group of related polymers showing a great variability in their structures according to their origin and cytological localization (Table 1, Fig. 1).

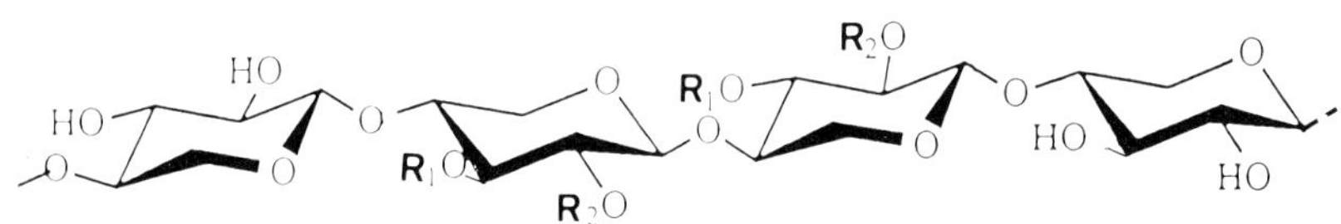

Fig. 1. Xylan main chain and substitution patterns. R1 and R2 may vary according to data in Table 1.

On the basis of the substituents, three main families of xylans can be considered. The arabinoxylans from cereal and grasses (519) with more complex arabinosyl oligosaccharide side chains, the true glucuronoxylans found in Gramineae and hardwoods, and the glucuranoarabinoxylans typically found in Gymnosperms but also in dicotyledons (459, 488, 489) and, with additional substituents of β-D-galactose, in several perennial plants (247, 518).

Glucomannans and galactoglucomannans

Glucomannans and galactoglucomannans are hemicelluloses typically found in softwood, in which they account for about 30 percent of the dry weight. They are often improperly referred to as "mannans" (157, 164), a misleading designation because it does not describe their true chemical structure. In glucomannans, glucose and mannose are integral parts of the main chain; the term "mannan" should be used only for true mannans such as yeast or seed mannans. Glucomannans consist of a chain where 1,4-linked β-D-glucopyranose and β–D-mannopyranose residues alternate irregularly in the main backbone. In galacto-glucomannans, terminal D-galactopyranosyl substituents are attached at O-6 of either mannose or glucose (Fig. 2, Table 1).

Glucomannans are found both in hardwoods and softwoods, but they account for only 3-5% of the hardwoods' dry weight (Table 1). In softwood, mannose-containing hemicelluloses may represent more than 20% of the wood, and like xylans, show some degree of polymolecularity (147). In all softwood hemicelluloses, acetyl substituents are important structural elements.

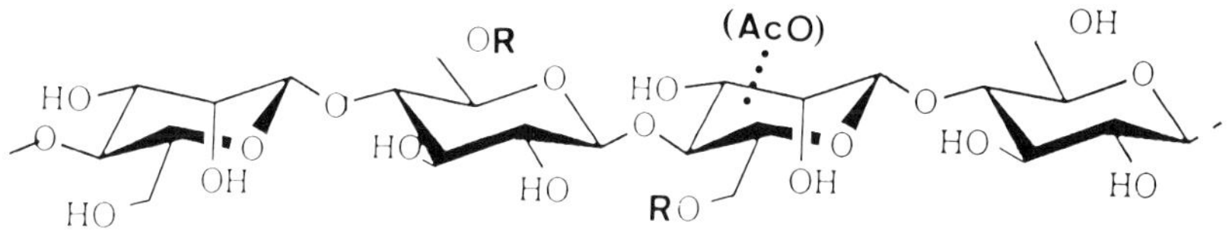

Fig. 2. Glucomannan main chain. R: β–D-galactopyranosyl substituents; Ac: acetyl group.

INTERACTION WITH OTHER WALL COMPONENTS

In the plant cell wall, cellulose microfibrils serve as a support onto which the other wall polymers interact through non covalent secondary forces. Covalent interactions occur between hemicelluloses and other non-cellulosic polysaccharides as well as with lignin and lignin-like compounds. These interactions give the plant cell walls their mechanical resistance and are also responsible in great part for the higher resistance of cellulose and hemicelluloses to enzymatic degradation.

Non covalent interactions

Xylans form extended, β-1,4-linked polysaccharide chains in which the internal units are linked through diequatorial 1e-4e

Table 1. Major hemicelluloses in hardwoods and softwoods.

Wood	Hemicelluloses	Usual proportion (% dry wood)	Main chain	Side chain and substituents	Relative proportion (%)
Hardwoods	Glucuronoxylan	15-22	→4)-Xyl*p*(1→4)-Xyl*p*-(1→	4-OMe-Glc*p*A-(1→2)	10
				4[Xyl*p*]$_{x}$ (1→3)	2-3
				O-Acetyl	4-7
	Glucomannan	3-5	→4)-[Man*p*-(1→4)]$_{2\text{-}3}$-Glc*p*-(1→4)-Man*p*		
Softwoods	Arabinoglucuronoxylan	4-10	→4)Xyl*p*-(1→4)-Xyl*p*-(1→	Glc*p*A-(1→2)	10-16
				4-OMeGlc*p*A-(1→2)	
				Ara*f*(1→3)	6-10
	Glucomannan	10-15	→4-[Man*p*-(1→4)]$_{3\text{-}4}$Glc*p*-(1→4)-Man*p*	Gal*p*(1→6)	≤3
				O-Acetyl	30
	Galactoglucomannan	3-8	→4-[Man*p*-(1→4)]$_{3\text{-}4}$-Glc*p*-(1→4)-Man*p*-	Gal*p*-(1→6)	30
				O-Acetyl	30

linkages. With the xylopyranosyl ring in a ${}^4C_1$ conformation, the preferred shape of xylans is a slowly twisting ribbon with a monomer repeat at 0.5 nm (14). Relative humidity affects the degree of crystallinity. Dea and Rees (132) have shown that the role of xylans in non-covalent interactions with cellulose is highly dependent on its hydration state when coating the microfibrils (348). Because xylose lacks the hydroxymethyl group of the hexoses, it cannot form a network of intra- and inter-chain hydrogen bonds like that of cellulose crystals (14). The formation of stable associations of xylan with cellulose depends on their conformation within the cell wall (334). In this respect, the substitution of the main chain with other sugar units and with acetyl is important, as is also the hydration of the cell wall for the formation of hydrates (348).

Unlike the pure mannans, glucomannans and galacto-glucomannans from wood are not crystalline in their native form. The acetyl substituents apparently interfere and prevent the regular orientation of the chains (251). The crystallization behaviour of alkali-extracted glucomannans from various origins and their association with cellulose has been studied by Chanzy et al. (103). The affinity of glucomannan for cellulose could always be established.

Covalent interactions

Covalent interactions essentially involve lignin and lignin-like compounds. The existence of covalent lignin-polysaccharide complexes in the plant cell wall remained undescribed for several decades. The use of model compounds and selective chemical and enzymatic degradations demonstrated the presence of covalent, ether (171, 187), ester (225) and glycosidic linkages (248, 277). With the exception of the glycosidic bond, lignin-carbohydrate linkages are also alkali labile. The best documented demonstration of covalent interactions refer to the benzyl ester bonds formed between the carboxyl of the 4-O-methyl glucuronic acid of xylan and lignin. Clear evidence of ether linkages was also obtained. Using the selective oxidation with DDQ (2,3-dichloro-5,6-dicyanobenzoquinone), Watanabe et al. (511) demonstrated that softwood glucomannans were etherified with benzyl positions of lignin, preferentially at O-6 of mannose and glucose, whereas arabinoglucuronoxylans were etherified through position -2 and -3 of the xylosyl units.

ULTRASTRUCTURAL LOCALIZATION

In the cell wall, cellulose fibers are embedded in a matrix of hemicelluloses and lignin. The relative distribution of the different polymers within the cell wall at the microscopic level largely influences the susceptibility to enzymatic degradation of each constituent. Recent developments in the cytochemical markers of the wall polymers have brought a better understanding of the ultrastructural relationship between cellulose, hemicelluloses and lignin (352, 420).

Takabe et al. (484) studied the correlations among chemical analysis, transmission electron microscopy and isotope tracer technique. Cellulose in conifer tracheids was mainly deposited in the middle of the S_2 layer, whereas xylans and glucomannans were more concentrated in S_1, outer S_2, and S_3 layers. In xylem of differentiating trees, cell wall polysaccharides are produced before lignification starts (485). The use of specific labelling of xylans and glucomannans with xylanase and mannanase-gold complexes, respectively, revealed that the distribution of hemicelluloses in the cell walls was heterogeneous (420) although the arrangement of hemicelluloses and lignin around cellulose microfibrils pointed to a lamellar organization. This agrees with the results obtained by scanning transmission electron microscopy (419) and by Raman spectroscopy (12), both demonstrating the orientation of polysaccharides and lignin in the cell wall. All these results confirm the associations that exist between all the constituents of the lignified cell wall.

Recently, enzyme-gold probes were used to label cellulose (39, 49, 77), xylans (420, 505), and glucomannans (420). A good labelling of the entire wall could be achieved with such probes. Recent findings of an identical cellulose binding domain (CBD) in the molecular structure of cellulases and some xylanases (185), however, make the specificity of cellulase and xylanase-gold probes for their respective substrate questionable.

MECHANISMS OF DEGRADATION OF POLYSACCHARIDES BY FUNGI

The cell wall is the first barrier to be overcome by fungi during the infection and colonisation process. The colonisation results in a more or less intense degradation of the cell wall components. The wood rotting fungi can alter the cell walls either *simultaneously*, when they degrade at the same time lignin and polysaccharides, or *selectively*, when the attack results in the preferential removal of one or more components (157). Cellulose and hemicellulose breakdown by fungi is the result of the synergystic action of several, primarily hydrolytic enzymes, but a few oxidative enzymes are also involved in cellulose degradation. In addition, non enzymatic oxidative systems, principally activated oxygen, may contribute to polysaccharide breakdown during fungal attack.

Cellulolytic enzymes

Fungal cellulases are traditionally classified into three major categories according to their mode of hydrolysis of substrates:

- Endo-1,4-β–glucanase [EG; EC 3.2.1.4]
- Exo-cellobiohydrolase [CBH; EC 3.2.1.91]
- β–glucosidase [EC 3.2.1.21].

Endo-1,4-β–glucanase splits glycosidic linkages in the cellulose chain by random internal cleavage, releasing fragments whose nonreducing ends become substrates for the exoenzymes. The microbial hydrolysis of cellulose is a complex process that involves various combinations of endo- and exoglucanases which vary according to the microorganism (524). In all these systems, however, cellobiohydrolases seem to be key enzymes with a cooperative action, as demonstrated for CBHI and CBHII with one or more endoglucanases and at least one β–glucosidase (523). During the degradation of native cellulose within the network of the wood cell wall, both endohydrolases and cellobiohydrolases tightly adsorb on the cellulose surface. Many cellulases possess a CBD which is separated from the catalytic domain by a linker sequence (356). The CBD has also been demonstrated in bacterial cellulases. Gene fragments from the endoglucanase of *Cellulomonas fini* were cloned (141). The corresponding polypeptides exhibited the same affinity for cellulose as the enzymes, but had no hydrolytic activity. In addition to acting as an anchor for the catalytic domain, CBD plays a role in the disruption of the cellulose fibre structure by exfoliating microfibrils, releasing the ends of cellulose chains and causing roughening of the surface (141). This mode of action of CBD in disrupting cellulose microfibrils is possibly crucial in the first steps of fibril opening by fungi.

Hemicellulolytic enzymes

Xylanolytic enzymes are active in the digestion of xylans. The solubilization of xylans by fungi involves random internal cleavages in the β–1,4-xylosyl backbone. This is achieved by the action of endo-1,4-β-D-xylanases (EC 3.2.1.8). A complete degradation, however, involves the synergistic action of a set of enzymes (52, 157, 395) including 1,4-β-D-xylosidases (EC 3.2.1.37) which hydrolyze xylo-oligosaccharides released by endoxylanases into xylose, i.e. α–L-arabinofuranosidase (EC 3.2.1.55) in the case of arabinoxylans; α-D-glucuronidase (EC 3.2.1) in the case of glucuronoxylans, and acetyl esterase (EC 3.1-1.6). All activities seem to be either constitutive or induced when the fungi are growing on appropriate substrates. In this respect filamentous fungi, which are good wood degraders, produce a broad spectrum of activities against xylans. Recombinant DNA techniques have allowed cloning of genes coding for the various enzymes that degrade heteroxylans (214). Comparison of sequences has revealed domains that are common to both cellulases and xylanases (185). Each of the enzymes from *Pseudomonas fluorescens* involved in xylan hydrolysis has a bifunctional organization including a CBD in which the binding to cellulose is mediated by a highly homologous conserved domain of about 100 amino acids connected to the rest of the protein by linkers (185).

In addition to hydrolysing enzymes, extracellular cellulose-oxidizing enzymes contribute to cellulose breakdown by some fungi. For example, quinone oxidoreductase (CBQ, EC 1.1.5.1) found in *Sporotrichum pulverulentum* (513) is a flavoprotein which requires a quinone to oxidize cellobiose to cellobiose-δ-lactone. Another extracellular cellobiose oxidizing enzyme produced by numerous cellulolytic fungi, cellobiose oxidase, (CBO) is a heme protein which, like CBQ, contains the coenzyme FAD as a prosthetic group. This enzyme has been demonstrated to enhance crystalline cellulose degradation by cellulases (25). In both enzymes the reduction of FAD is considered to couple directly to the oxidation of cellobiose (431). Enzymes that contribute to cellulose degradation, like cellulases, bind strongly to microcrystalline cellulose and this has suggested that they may contain a cellulose-binding domain (401). CBQ is involved in lignin, cellulose, and hemicelluloses degradation. Hypothetical roles for this enzyme which is intermediate between lignolytic and polysaccharide degrading enzymes have been suggested by Eriksson et al. (157). The concentration of H_2O_2 regulates the action of CBO either on lignin or cellulose degradation (25).

Non enzymatic agents involved in cellulose degradation

The difficulty in penetrating the plant cell wall network limits the efficiency of hemicellulases and cellulases in hydrolyzing polysaccharides within the lignocellulosic matrix. This limitation is essentially due to their size, relative to the pore size of the wood cell wall (168). Small-size agents can be effective during the first steps of polysaccharide degradation. Kirk et al. (257) have shown that samples of cotton cellulose depolymerized by a brown-rot fungus are similar to those decayed by Fenton's reagent. In both cases the molecular size of the degraded cellulose was indicative of depolymerization due to cleavage in the non crystalline regions. These finding are in good agreement with the observation that cell wall lysis by brown-rot fungi occurs at some distance from the hypha (193).

A generally accepted hypothesis on the nature of cellulose depolymerizing agents postulates the role of oxygen-derived radicals. These can be generated as •OH radical by Fenton's reaction from Fe(II) and H_2O_2, or as the superoxide radical ($O_2^{\cdot-}$). H_2O_2 plays an active role for the release of crystalline cellulose degrading agents.

Hydroxyl radical formation during plant-fungus interaction may occur in different ways. The enzymes involved in cellulose and in lignin breakdown may be the indirect source of non enzymatic agents also involved in cellulose degradation. For instance, CBO can generate superoxide anions (15), and cooperate with hydrogen peroxide and ferric ions to form hydroxyl radicals (280). Lignin peroxidase (LiP) can produce hydroxyl radicals by a mechanism which includes secondary metabolites, veratryl alcohol and oxalate (26). LiP can also catalyze the reduction of molecular oxygen to produce superoxide in the presence of H_2O_2, and also superoxide radical ($O_2^{\cdot-}$) (394). Oxalic acid has been suggested to reduce Fe(III) to Fe(II) and thus to be indirectly involved in cellulose breakdown (158). Hypotheses concerning the acid-mediated depolymerization of cellulose and hemicelluloses by oxalic acid have been recently discussed (194, 450).

VISUALIZATION OF POLYSACCHARIDE–DEGRADING ENZYMES DURING FUNGAL ATTACK

Transmission electron microscopy provides some understanding of the biochemical events whereby the cell wall polymers are modified or degraded during fungal invasion of plant tissues (64).

Immunogold labelling of xylanases of *P. chrysosporium* has demonstrated that these enzymes, associated to the fungus cell wall, are present at an early stage of hyphal development (246). This suggests that a part of the xylanases present in the fungus is constitutive.

Immuno-scanning electron microscopy using monoclonal antibodies against xylanases from *Postia placenta* suggested that the enzymes could be located in the extracellular sheath, on the hyphal surface and within the soluble extracellular material (193). These observations suggest that xylanases are located at different places depending on the physiological stage of the fungus, being bound to the hyphal wall and thus active during primary growth or associated with the extracellular glucan polysaccharide during idophasic metabolism.

Immunocytochemistry demonstrated that β–glucosidase from *Trichoderma reesei* was associated to the outermost exopolysaccharide layer and to the plasma membrane (464). Using monoclonal antibodies, Nieves et al. (349) demonstrated that cellobiohydrolase (CBH-I) from *T. reesei* binds preferentially to ordered cellulose and endoglucanase (EG-I) in the amorphous regions. The immunoprobes labelling CBH-I and EG-I were on the same optical plane, thus leading the authors to the conclusion that EG cleaves the chain and is immediately followed by CBH-I, which binds onto the same site. Localization by immuno-scanning electron microscopy of extracellular wood-degrading enzymes excreted by *Postia placenta* revealed that extracellular xylanases and cellulases were weakly bound to the hyphal sheath (193). This suggests a definite role of the extracellular sheath during fungal degradation of woody substrates (Nicole et al., this volume).

Electron microscopy observations of the patterns of degradation of the substrates attacked by microorganisms with different sets of enzymes (418) have allowed a better understanding of the sequence of action of the enzymes that degrade lignocellulosic materials. Correlations between data from enzymology, molecular biology and electron microscopy will enable us to better understand the mechanisms of degradation of the major cell wall polymers by microorganisms.

LIGNIN BIODEGRADATION IN CELL WALLS OF WOODY PLANTS

Robert A. BLANCHETTE

Department of Plant Pathology, University of Minnesota, St. Paul, MN 55108, USA

Terrestrial vascular plants contain considerable quantities of lignin, an amorphous phenylpropanoid structural polymer that binds cell wall components and cell walls together providing structural stability to plants. Lignin lacks the regular and ordered repeating units found in other natural polymers (293). Instead, it is a composite of physically and chemically heterogeneous materials (9, 157, 351, 434). The three phenylpropanoid monomers that give rise to lignin include coniferyl, sinapyl and p-coumaryl alcohols. Most of the lignin in gymnosperms (more than 95%) is guaiacyl lignin since it is derived from coniferyl alcohol, whereas in angiosperms a combination of guaiacyl and syringyl lignins occur and are composed by varying ratios of coniferyl and sinapyl alcohol derived units (293). In normal wood of gymnosperms, lignin content ranges from 24 to 33% and in temperate angiosperms between 16-25% (434, 490). In compression wood, the reaction tissue of gymnosperms that forms on the lower side of leaning trunks or branches, lignin content varies between 35-40%. Tension wood, formed on the upper side of leaning stems and branches of angiosperms, contain only 15-20% lignin (69, 490).

Lignin occurs in the compound middle lamella as well as in the secondary wall. Since the secondary wall constitutes the largest proportion of the total cell volume, most lignin is located in this region. The greatest concentration, however, is in the compound middle lamella. Variations in concentration and composition of lignin occur among different types of cells within a particular wood species, and also within wall layers of specific cells. For example, the cell walls of vessel elements from *Acer*, *Betula*, *Tilia* and other angiosperms (Fig. 1A) are morphologically different than fibers and contain greater concentrations of lignin with a larger number of guaiacylpropane units as compared to fiber walls (68, 157, 355). In gymnosperms (Fig. 1B), differences in lignification exist between earlywood and latewood tracheids, and in some *Pinus* species, ray parenchyma cells of sapwood are free lignin (24, 143, 165).

Lignin is often considered to be a primary factor that imparts resistance to microbial degradation of wood. The intimate spatial association of lignin and polysaccarides within the woody cell wall provides a protective barrier that impedes degradation of cellulose. Once lignin surrounding cellulose fibrils is removed or modified, cellulose is more accessible to microbial enzymes and may be effectively degraded. Lignification in many living plants also can be a mechanism for disease resistance (499), and is produced as a defense mechanism to pathogenic fungi or in wound response (56, 63).

LIGNIN DETERMINATION

Lignin can be detected with a variety of chemical analyses. Acid hydrolysis is widely used to obtain a gross estimate of lignin content by determining the acid-insoluble fraction (134, 148). Nitrobenzene and cupric oxide oxidations as well as thioacidolysis can be used to characterize the type of lignin (101).
For morphological studies, histological stains, ultrastructural methods, and spectrophotometric analyses are available. Color-forming reagents, such as phloroglucinol-hydrochloric acid and the Maule color reaction are routinely used in light microscopy (241, 337). Qualitative and semiquantitative estimates of lignin within cell walls with transmission electron microscopy are possible using potassium permanganate as a stain (70, 71). Increased electron density with $KMnO_4$ corresponds to areas of greatest lignin concentration (Fig. 1). Quantitative assays include ultraviolet microscopy (165, 172), interference microscopy (142), and an electron microscopy method that involves bromination of wood followed by detection of brominated lignin with Energy Dispersive X-ray Analysis (EDXA) (429, 430). Measurements of bromine concentrations across a cell wall provide a precise method for lignin localization in various cell wall layers (Fig. 2). Another electron microscopy method using EDXA and mercurization has also been recently described (514).

FUNGI THAT DEGRADE LIGNIN

Many different microorganisms can modify and/or degrade lignin, but white rot fungi are one of the most efficient groups known to metabolize this polymer (62, 255). This category, defined after the type of decay produced in wood, contains hundreds of species in the Basidiomycotina. All are able to degrade lignin, cellulose and hemicellulose from wood, but the rate and extent of degra-

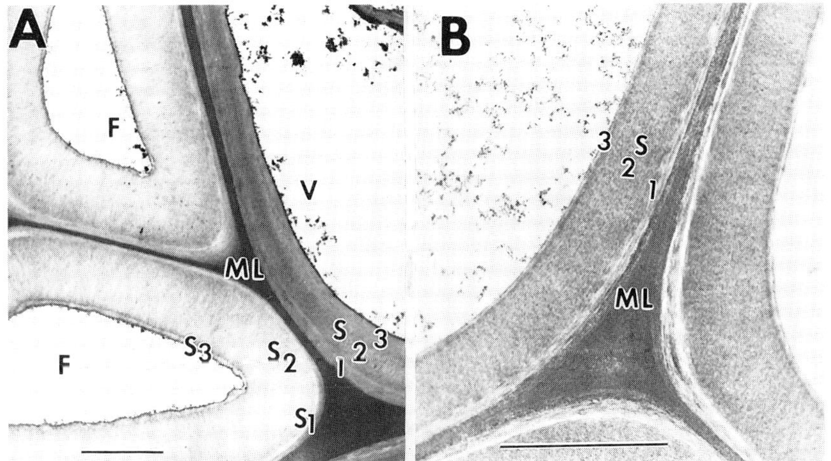

Fig. 1. Transmission electron micrographs of transverse sections from *Betula* (**A**) and *Pinus* (**B**). **A**, Cell walls of vessel and fibers show differences in secondary wall structure. **B**, Tracheids showing cell wall layers. $KMnO_4$ fixation allows areas with greater concentrations of lignin to appear more electron dense. Lignin distribution is evident throughout the cell walls but is more concentrated in the compound middle lamellae. ML = middle lamella, S_1, S_2, and S_2 = secondary wall layers, V = vessel, F = fiber. Bar = 3 μm.

dation of each cell wall component varies considerably. White rot fungi are usually classified into subcategories based on the simultaneous degradation of lignin along with cell wall polysaccharides as opposed to selective lignin degradation (62). Attempts to further classify common characteristics among white rot fungi have accommodated many species providing similar decay patterns in the same group, however, the extremely large variability observed makes it difficult to place many species into one of the groups. In addition to species of white rot fungi that preferentially degrade lignin or those that simultaneously attack all cell wall components, others produce both types of decay within the same substrate. Moreover, some species selectively digest lignin during incipient stages of decay but subsequently remove the residual cellulose in later stages of attack (157). The substrate also may alter the way in which lignin is degraded. In a study of decay in two tree species from the temperate rain forests of Chile, *Ganoderma australe* caused a highly selective degradation of lignin in *Nothofagus dombeyi*, a tree with one of the highest ratios of syringyl to guaiacyl lignin (S/G=6.3) whereas in *Laurelia phillipiana* (S/G=0.8) the degradation of all cell wall components was not selective (4). Genetic differences among strains of certain white-rot fungi also appear responsible for a high degree of variability in wood decay and lignin removal (380). Strains of *Phanerochaete chrysosporium*, such as BKM-F-1767, are excellent delignifiers, whereas others do not degrade appreciable amounts of lignin in comparison to the concentration of cellulose digested (67). In addition to these sources of variation, environmental factors (e.g. oxygen concentration, temperature, etc.) can have significant effects on the amount of lignin degraded by a given isolate (400).

Soft-rot fungi include a large number of species in the Ascomycotina and non-basidiomycetous Deuteromycotina that also can degrade lignin. These organisms produce cavities within the secondary walls (Type I) or erode cell wall layers (Type II) in a manner somewhat similar to white rot fungi. Significant losses of lignin may occur in angiosperms where secondary walls can be completely degraded (157). In gymnosperms, the attack is limited and soft rot cavities are the major form of degradation (350). Results using the nitrobenzene oxidation technique indicate that soft rot fungi preferentially degrade the syringylpropane units in lignin (350). Although guaiacylpropane units resist degradation, substantial losses of lignin can occur in various gymnosperms over extended periods of exposure to cavity-forming species of soft rot

---

Fig. 2. Transverse sections of *Betula* fixed with $KMnO_4$ (**A, C, E,** and **G**) and X-ray microanalysis of bromine distribution profile across cell walls after bromination treatment (**B, D, F, H**). Dotted lines show representative area where scan line was made on sections of brominated wood. **A, B** Sound wood showing bromine distribution profile and quantity of lignin in secondary wall and compound middle lamella. Background levels of bromine represented in lumen (L) are low. Within secondary walls (S) , the concentration of lignin remains relatively constant until the middle lamella region (ml) which contains the greatest concentrations. **C, D** Decay by *Trametes versicolor* showing an erosion of the cell wall near hypha (hy). The depletion of lignin before the eroded zone can be seen as an electron lucid zone in the cell wall and by the reduced bromine distribution profile. Increases were also observed at the location of hyphae (hy) indicating that some bromine reacted with the fungus. Localized delignification zones were evident within the cell wall. **E, F** Incipient stage of decay by *Phlebia tremellosa* exhibiting extensive delignification throughout the secondary wall near hypha (hy). Normal distribution of lignin was observed in adjacent cell that appeared unaltered and without hypha in cell lumen. Arrow indicates middle lamella region. **G, H** Cells delignified by *Phellinus pini* after advanced decay. Extensive delignification throughout secondary walls and middle lamella was evident by the lack of electron density in $KMnO_4$–treated samples and the low concentrations of bromine observed throughout the cell walls in brominated samples. Arrow indicates middle lamella region in bromine distribution profile. Dotted line = approximately 5-8 µm. (from 64).

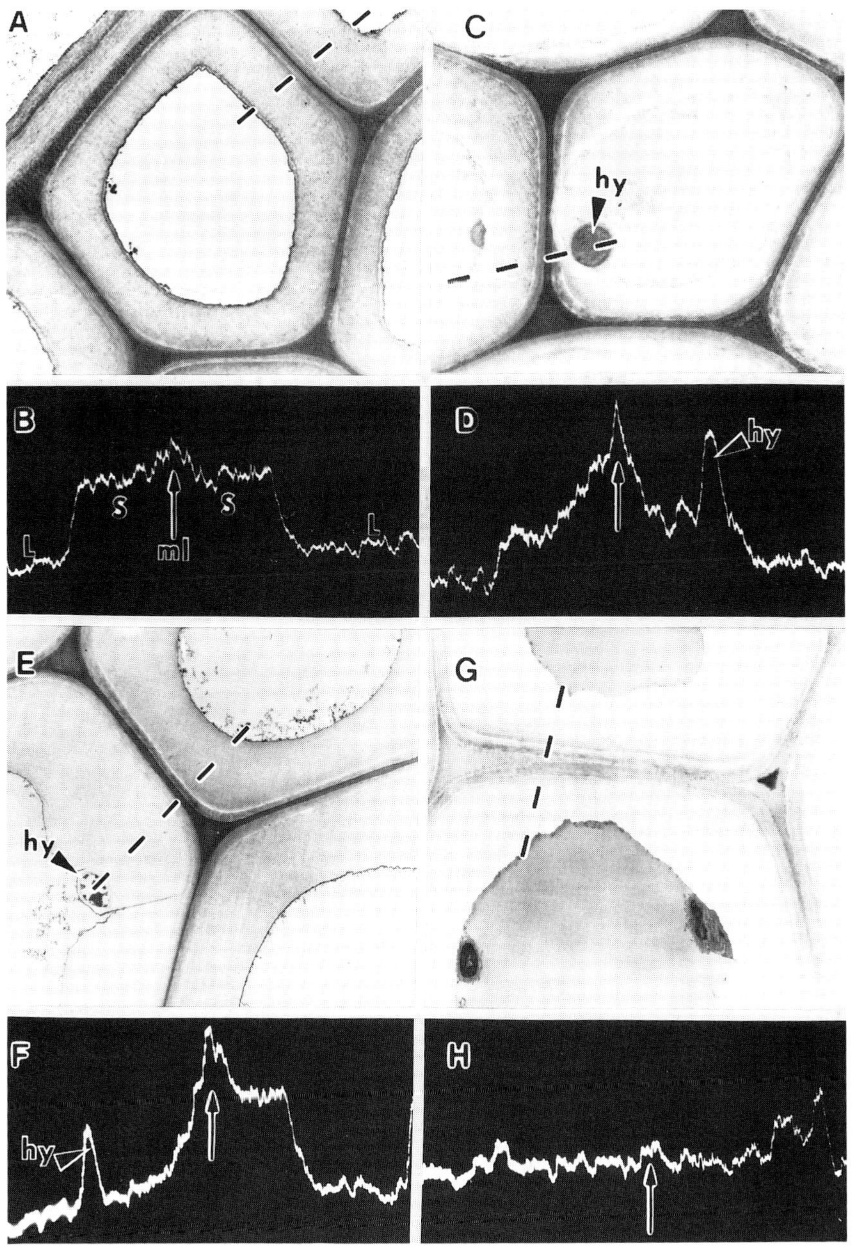
A
C
hy
B
S
ml
S
L
L
D
hy
E
hy
G
F
hy
H

fungi. Another large group of wood-destroying fungi cause brown rot. Agents of brown rot are Basidiomycotina that rapidly depolymerize cellulose. In advanced stages of decay, polysaccharides are extensively degraded, leaving a lignin-rich residue. The high concentration of lignin remaining in the decayed wood is evidence for a limited capacity for lignin removal. The lignin polymer is chemically altered, however, and several investigations have revealed that some degradation of lignin does inevitably occur (155, 222, 236).

LIGNIN DEGRADATION IN CELL WALLS

Morphology

Ultrastructural localization of lignin using $KMnO_4$ staining and the bromination-TEM-EDXA technique provides an accurate method for investigating how lignin is degraded from the cell wall by white-rot fungi. The electron dense staining pattern after $KMnO_4$ treatment allows visualization of relative lignin concentrations throughout the cell wall, whereas the bromination method provides a more precise determination of lignin in specific areas (Figs. 2A, B). Areas void of lignin have reduced electron density after $KMnO_4$ staining and have fewer bromine counts after bromination and X-ray analyses. In transverse sections of wood decayed by *Trametes versicolor*, a simultaneous attack of all cell wall components occurs and cell wall erosion takes place in the secondary wall adjacent to the lumen. The decay progresses from the lumen toward the middle lamella. Before cell wall erosion is evident, an electron-lucid region can be observed within a relatively narrow zone in the secondary wall (Fig. 2C). This zone reacts poorly to bromine and is free of appreciable concentrations of lignin (Fig. 2D). Observations carried out during various stages of cell wall erosion indicate that lignin removal precedes cellulose degradation (3, 70, 157). The zone of delignification by *Trametes versicolor* does not extend deeply into the cell wall, and it appears that the onset of cellulolytic activity immediately follows lignin removal. Progressively, lignin degrading agents move into the wall and complete erosion of the secondary cell wall takes place. Once the erosion reaches the middle lamella, this region is also degraded in a localized area. Similar patterns of secondary cell wall degradation occur by soft rot fungi that erode cell walls (Type II attack), but the compound middle lamellae are not degraded (157, 350).

Lignin degradation patterns are different for many species of white rot fungi that selectively remove lignin and do not extensively degrade cellulose. A diffuse attack resulting in large zones of delignification occurs within the cell walls. Even during an early stage of decay, lignin is depleted throughout the secondary wall (Figs. 2E, F). Once the delignification process has reached the compound middle lamella, the area between cells, and subsequently the cell corner regions are degraded . In advanced stages of decay, the secondary walls and middle lamellae appear completely delignified (Figs. 2G, H). Fungi that produce this type of decay differ from other white-rot fungi in the extremely diffuse nature of lignin removal that they initiate, as well as in the repression of the cellulolytic enzyme system.

Many factors influence lignin degradation, but even the most resistant substrate may be delignified. For instance, cells of compression wood are morphologically different than normal wood cells, contain high concentrations of lignin, possess more p-hydroxyphenyl units and show a more condensed lignin structure (69, 490). The delignification process caused by fungi that are adapted to colonizing gymnosperms follows patterns similar to those that occur in syringyl-rich angiosperms; a diffuse removal of lignin from the cell wall without substantial losses of cellulose occurs (Figs. 3A, B). The rate of delignification, however, is limited in compression wood (69). This appears primarily due to the high concentration of lignin within compression wood as well as to the altered composition of the lignin.

The extensive removal of lignin from an entire cell or group of cells originating from just a single hypha in the cell lumen is an intriguing aspect of the delignification process. The compounds responsible for lignin degradation must be small enough to diffuse into the cell wall through existing pores or inter-fibrillar spaces. Previous investigations have shown that wood decay enzymes cannot penetrate into walls of sound wood cells (65, 344, 465). Changes in pore size by non-enzymatic agents or depolymerization of a wall component such as hemicellulose could provide lignin modifying or degrading enzymes access into the wall. In a recent study of decay in tension wood, the ability of wood decay enzymes to diffuse through the gelatinous (G) layer and cause a simultaneous attack of the underlying secondary wall was demonstrated (69). The G layer consists entirely of crystalline cellulose that lines the cell wall with a relatively thick layer. Degradative enzymes from hyphae of *Trametes versicolor* moved through the layer and subsequently degraded all secondary wall components. Erosion troughs were evident beneath the gelatinous layers (Figs. 3C, D). In advanced stages of decay, the entire secondary wall and compound middle lamella

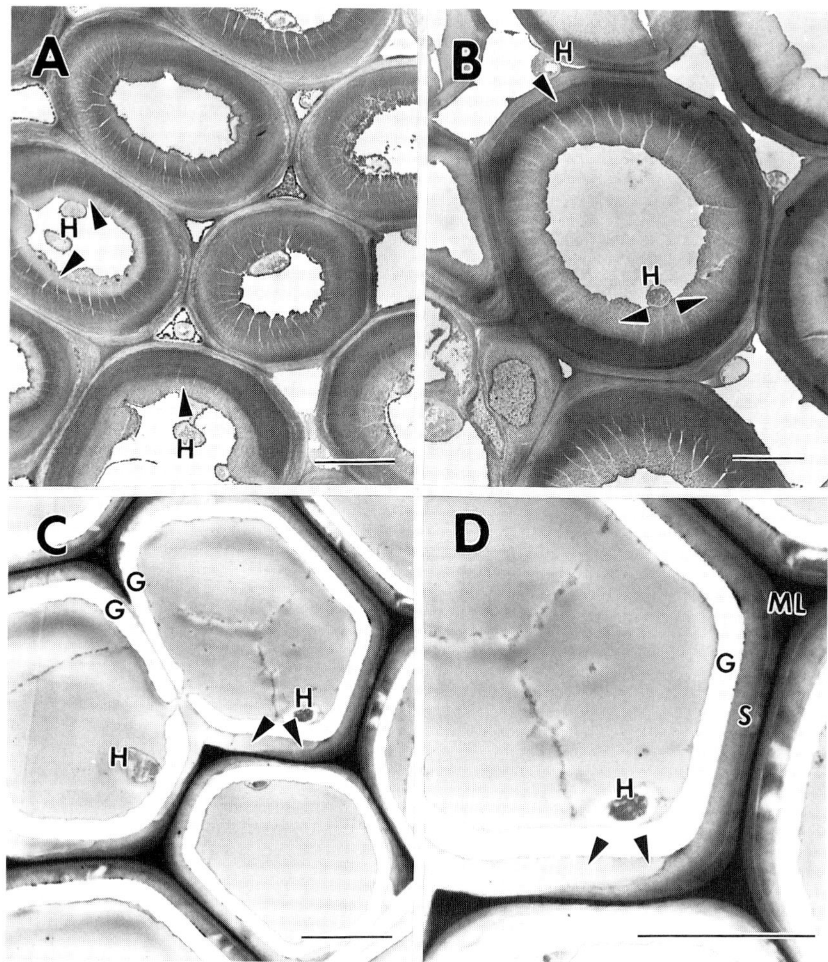

Fig. 3. Transmission electron micrograph of transverse sections from compression wood of *Pinus* (**A** and **B**) and tension wood of *Populus* (**C** and **D**). **A** and **B** Delignification of tracheids by *Syctinostroma galactinum* showing electron lucid zones in thick secondary walls of compression wood cells (arrowheads). Hyphae (H) are located in cell lumina, as well as within intercellular spaces in cell corners. **C** and **D** Decay by *Trametes versicolor* resulting in a complete erosion (arrowheads) of secondary wall layers beneath the gelatinous layer (G). No erosion was evident in the gelatinous layer but a full complement of wood degrading enzymes passed through this layer to incite a typical simultaneous decay of the underlying secondary wall and middle lamella. H = hypha, S = secondary wall, ML = middle lamella. Bar = 5 μm.

were degraded but the G layer remained relatively unaltered. Apparently a full complement of degradative enzymes diffused through the wall to degrade lignin, hemicellulose and cellulose. Although ultrastructural changes were minimal in the G layer, alterations were induced resulting in pores of size sufficient to allow the penetration even of cellulases with high molecular weights. Changes in cell wall porosity during incipient stages of white rot have also been demonstrated using cytochemical stains in wood chips pretreated with *Ceriporiopsis subvermispora* (66).

Biochemical and immunocytological aspects

The biochemical processes responsible for lignin degradation by white-rot fungi are not fully understood, but considerable progress has been made to determine the primary agents involved in ligninolysis. It is clear from previous investigations that the fungal extracellular system is oxidative and nonspecific (187, 255). A great deal of research has concentrated on elucidating the role of lignin peroxidase (an enzyme which can oxidize veratryl alcohol in the presence of hydrogen peroxide) during lignin metabolism (205, 224, 255). Using model compounds, peroxidase has been shown to oxidize the aromatic nuclei of some substructures of lignin by one electron, thus producing radicals that undergo a variety of nonenzymatic substitutions and spontaneous degradative reactions (256). Subsequent nonspecific oxidation reactions follow, causing ring cleavage and producing a variety of intermediate degradation products. This nonspecific oxidation process has been referred to as "enzymatic combustion" (255). The oxidative potential of the lignin substrate components influences oxidation rates. For the various types of lignin, the oxidative potential of syringyl would be greater than that of guaiacyl, which in turn is higher than that of p-hydroxyphenyl lignin (255). This is in agreement with the sequence observed in various types of wood decayed by fungi in natural systems. Syringyl lignin is degraded more rapidly than guaiacyl, followed by p-hydroxyphenyl lignin which is the most resistant (68, 69, 163). The decay patterns observed for soft rot fungi also have been postulated to be caused by some oxidase activity. An oxidase with reduced oxidative potential produced by soft rot fungi may explain the degradation of syringyl-rich areas, such as the secondary wall, while the guaiacyl-rich middle lamellae is left unaltered (350).

There are many white-rot fungi that are aggressive lignin degraders, but to date have not been found to produce lignin peroxidase. The white rot fungi, *Ceriporiopsis subvermispora*, *Dichomitus squalens*, *Lentinula edodes*, *Phlebia brevispora* and *Rigidoporus lignosus* produce Mn-peroxidase, laccase and other enzymes but not lignin peroxidase (176, 288, 374, 424). These results imply that lignin peroxidase is not essential for lignin biodegradation. Mn-peroxidase is produced by all these fungi. In vitro depolymerization of synthetic lignin has been demonstrated for Mn-peroxidase mediated reactions, suggesting that this enzyme has an important role in ligninolysis (510). Mn-peroxidase oxidizes Mn(II) to Mn(III), which may act as a highly mobile oxidant. Mn(III) also is of sufficiently small size to readily move into the cell wall, generating radicals deep within the lignocellulose matrix (170). Once bonds are broken and the cell wall is "loosened", low- and high-molecular weight enzymes could be able to move into the wall. A number of immunocytochemical studies have shown that various degradative enzymes are found in regions of the cell wall that have been altered by incipient decay and display a looser, more open fibrillar structure (65, 127, 422, 465). Other enzymes such as laccase and H_2O_2-generating enzymes also have possible roles in lignin degradation. These enzymes have been recently visualized within fungal hyphae and decayed wood (128, 344). Immunocytochemical studies using gold labelled antibodies of low molecular weight fungal siderophores detected the spatial distribution of chelators during decay by brown rot fungi (240). Similar investigations of transition metals and chelators produced by white rot fungi, alone and in combination with immunogold-labelling of degradative enzymes, should provide a better understanding of the early degradative events.

Other low molecular weight compounds, such as porphyrins, combined with various organic solvents, show similar activities as heme-containing peroxidases and also mimic the delignification process of wood (198, 365). An ultrastructural study of wood treated with various metalloporphyrins for 48 hours revealed degradation patterns indistinguishable from that caused by many weeks of fungal decay in selectively delignified wood (365).

In addition to lignin degradation, hemicelluloses are also depleted by biomimetic catalysts. In naturally decayed wood, loss of hemicelluloses always accompanies lignin removal (3, 62, 157). The oxidation of lignin has an effect on lignin-hemicellulose bonds and also appears to affect directly hemicelluloses. The presence of xylanase and other hemicellulases in wood during incipient stages of decay provide strong evidence that these enzymes may move into the secondary wall soon after the initial oxidation process begins (65, 422).

Although our understanding of lignin degradation from wood cell walls is far from complete, a consortium of enzymatic and non-enzymatic processes must be involved in fungal ligninolysis. One scenario of participating primary agents includes the involvement in the process of enzymatically produced Mn(III), of H_2O_2–generating enzymes, such as pyranose oxidase (128), glyoxal oxidase (253), and hemicellulases, as well as nonenzymatic polymer cleavage by cation radicals. These early events are accompanied or followed by other degradative enzymes such as lignin peroxidase and laccase that progressively degrade lignin to CO_2 and H_2O. Variation in decay patterns produced by white rot fungi are apparently due to different degradative agents produced and their sequence during the chain of events that are involved with lignin biodegradation from woody cell walls. The regulation of these enzyme and nonenzymatic mediated processes is an important aspect requiring additional investigation. Immunocytochemical and ultrastructural studies will help to determine which are the key agents during lignin degradation and also to establish the sequence of events that occurs in situ during diverse wood degradation processes.

FUNGUS–INDUCED DEGRADATION AND REINFORCEMENT OF DEFENSIVE BARRIERS OF PLANTS

Pappachan E. KOLATTUKUDY, Jörg KÄMPER, Ute KÄMPER, Luis GONZÁLEZ-CANDELÁS, Wenjin GUO

Ohio State Biotechnology Center, The Ohio State University, Columbus, Ohio 43210, USA

Fungal infection of host plants involves penetration of the host barriers by extracellular enzymes secreted by the invading pathogens. Fungal attack triggers a variety of defense reactions in the hosts including production of phytoalexins and antifungal proteins that degrade fungal walls or cause other deleterious effects on the invading pathogen. Another defense involves reinforcement of the host cell walls by producing cross links and by deposition of phenolics and aliphatics. In this paper, we shall briefly review the advances in our understanding of the basic mechanisms involved in fungal efforts to degrade the host barriers and the host's efforts to reinforce its walls to impede the ingress of the pathogen.

CUTICLE PENETRATION

Structure of cuticle

The plant cuticle is attached to the epidermal layer of cells via a pectinaceous layer. The cuticle is composed of an insoluble biopolyester, cutin, that is embedded in a complex mixture of hydrophobic materials collectively called wax.

Cutin monomers are composed mainly of the C_{16} and C_{18} families of hydroxy and hydroxy epoxy fatty acids (230, 261, 273) (Fig. 1). Indirect chemical studies indicated that most of the primary hydroxyl groups and about one half of the mid-chain hydroxyl groups are in ester linkages in the polymer (230, 261). More recently, high resolution ^{13}C NMR with cross-polarization magic-angle spinning (CP/MAS) as well as direct depolarization methods indicated that cutin contains CH_2 chains that have considerable motion and that secondary alcohol esters probably constitute rigid cross–links (472).

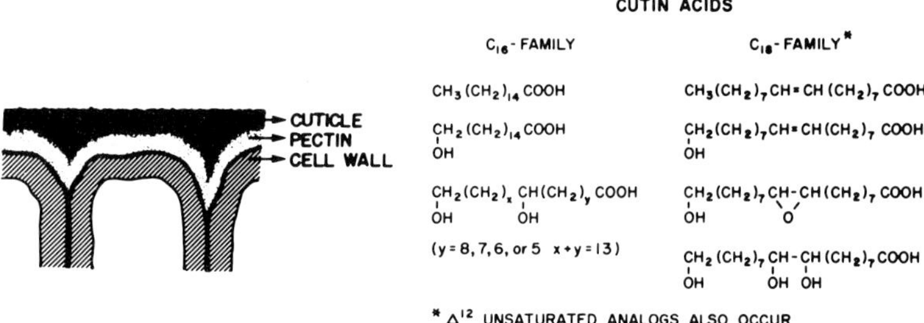

Fig. 1. Schematic representation of cuticle (left) and structure of the major monomers of cutin (right).

Role of cutinase in cuticular penetration

Cutin constitutes the major physical impediment for direct penetration of the infection peg into the host. Whether penetration through cutin is an enzyme-assisted process has been debated for a long time (265, 497). Cutinase, the postulated fungal enzyme involved in this process, was purified first from *Fusarium solani* f. sp. *pisi (Nectria haematococca)* in 1975 (396, 397) and the enzyme has been extensively studied (264). Recently, X-ray crystallography studies showed that the fungal lipases belong to a novel structural class of hydrolases that bridge lipases and esterases (310). Immunoelectron microscopic examination showed that cutinase is produced during infection of the host (442). If cutinase assists the penetration of the fungal pathogen through the host cuticle, the enzyme should be targeted at the penetrating point. In cases such as *F. solani* f. sp. *pisi* that involve direct penetration of the germinating spore through the host cuticle, the enzyme should be targeted to the germ tube. In cases where appressorium formation is involved, cutinase should be targeted at the infection peg that arises from the appressorium that penetrates the host. Immunofluorescence localization of cutinase in the germinating spores of *F. solani* f. sp. *pisi* and *Colletotrichum gloeosporioides* demonstrated this expected targeting (390). In several hosts/pathogen systems, inhibition of cutinase with chemicals including suicide inhibitors, polyclonal antibodies (139, 265, 269) and monoclonal antibodies (92) resulted in protection of hosts against attack by fungal pathogens (Table 1).

Cutinase inhibitor sprayed on papayas in the field protected the fruits from anthracnose (269). Reduced virulence of cutinase-deficient mutants of F. *solani* f. sp. *pisi* (129) and C. *gloeosporioides* (138) and restoration of virulence by the addition of exogenous cutinase were demonstrated. *Mycosphaerella* sp. that infects papaya fruits only through wounds acquired the ability to infect intact papaya fruits when a cutinase gene from *F. solani* f. sp. *pisi* was introduced in *Mycosphaerella* in a manner that the cutinase gene was inducible by cutin or cutin monomers. This infection could be pre-

Table 1. Plants shown to be protected against phytopathogens by cutinase inhibitor.

Host	Pathogen	Reference
Pea Stem	*Fusarium solani* f. sp. *pisi*	442
Papaya Fruits	*Colletotrichum gloeosporioides*	269
Apple Leaves	*Venturia inaequalis*	263, 276
Corn Leaves	*Colletotrichum graminicola*	263
Pepper Fruit	*Colletotrichum capsici*	a
Gerbera Flower	*Botrytis cinerea*	92

a: Ettinger and Kolattukudy, unpublished.

vented by antibodies prepared against the *Fusarium* cutinase (140). Gain of virulence by enhanced expression of cutinase was also observed in *F. solani* f. sp. *pisi* isolates. Isolate T-8, which has multiple cutinase genes and produces large amounts of cutinase (463), is highly virulent on intact pea stem but T-30, which has a weakly expressed cutinase gene, is not highly virulent on intact pea stem whereas both have equally high virulence on pea stem with a mechanically breached (pin prick) cuticle/wall barrier (275). When a cutinase gene from T-8 was transferred to T-30, transformants that expressed high levels of cutinase approaching that of T-8 also showed high virulence approaching that of T-8 (Table 2) (271). One transformant that produced less cutinase than the wild type T-30 also showed little virulence.

Two gene disruption studies were recently presented as evidence against any significant role of cutinase in virulence. A cutinase gene-disrupted *F. solani* f. sp. *pisi* transformant infected the host when assayed at one spore concentration (470). Our bioassays with the same gene-disrupted mutant showed drastically decreased virulence and microscopic examination of the infection areas just above the seed, a normal site of infection in the field (279), showed that the mutant penetrated through stomata. The wild type showed more generalized infection, although not as severe as T–8. Bioassays in which pea seeds were planted in vermiculite mixed with fungal spores showed that disruption of cutinase gene caused drastic decrease in virulence (Flaishman, Rogers and Kolattukudy, unpublished). Thus, the results obtained with this mutant support the conclusion that cutinase assists the highly virulent pathogens to penetrate the host cuticle.

Table 2. Cutinase activity and degree of virulence of strain T-30 transformed with strain T-8 cutinase gene.

F. solani f. sp. *pisi*	PNBase Activity (% of T–8)	Infection
T–30	25	20
Transformant 9	91	80
Transformant 19	43	70
Transformant 20	30	30
Transformant 5	0	10

PNBase: *p*-nitrophenyl butyrate hydrolase; from ref. 271.

In the other gene disruption study, a cutinase gene was disrupted in *Magnaporthe grisea* and this transformant showed virulence (483). Since the data in this paper showed that the transformant retained a major part of its ability to produce cutinolytic activity, this experiment cannot be used to assess the role of cutinase in virulence.

The relative importance of physical force and assistance by cutinase probably vary a great deal in the different host-pathogen systems. It is clear from the many lines of evidence that cutinase can play a significant role in infection by fungi. It is becoming increasingly clear that organisms that have retained the ability to infect their hosts have evolved ways to compensate for the loss of expression of a single gene. In the extensively studied case of *Erwinia* pectate lyases (119), certain genes are expressed only in the host (118) and host components induce expression of normally weakly expressed *pelA* (79). Animal pathogens are also known to express certain genes only while they are in the host (302). We found multiple cutinase genes in highly virulent *F. solani* f. sp. *pisi* isolates even by normal Southern hybridization; there may be others that have not been detected (265, 463).

The unique cutin monomers, produced by the cutinase carried by the virulent spores that land on plants, trigger transcriptional activation of the cutinase gene (265). Upon washing, only the highly virulent strains of *F. solani* f. sp. *pisi,* and not the weakly virulent strains released significant amount of cutinase. Upon contact with cutin, cutinase induction occurred only with the highly virulent strains which could readily infect intact pea stem. Thus, cutinase induction caused by the cutin monomers was correlated with virulence on intact host. This induction was found to be caused by the unique cutin monomers (522) and nuclear run on experiments showed that these monomers caused transcriptional activation of cutinase gene (271).

Identification of cutinase promoter elements and proteins that bind them

Plant cutin monomers trigger, and glucose suppresses, the expression of the cutinase gene of pathogenic fungi. That the promoter activity of the cutinase gene of *F. solani* f. sp. *pisi* resided within the 360 nucleotides immediately 5' to the cutinase initiation codon was demonstrated by transformation of *F. solani* f. sp. *pisi* and *C. capsici* (463). Transformation with vectors containing the 5' flanking region of the cutinase gene or its deletion mutants from *F. solani* f. sp. *pisi* fused with a chloramphenicol acetyltransferase *(CAT)* reporter gene and a constitutive promoter fused with a hygromycin phosphotransferase gene showed that *CAT* induction required a 360-bp, or longer, segment of the 5' flanking region of the cutinase gene (21). To elucidate further the nature of the promoter, a more detailed *Bal* 31 deletion analysis was done (249). A positive promoter element was found to be essential for high levels of *CAT* induction. Thus deletion and mutation of an *Sp1* type (mammalian) element, homologous to G-rich regions of the promoter sequences in *gpd*A and *pgk*A of *Aspergillus niger* and *A. nidulans,* caused loss of inducible expression of CAT. Nuclear proteins specifically bound to a 31 bp DNA fragment (–305 to –335) containing the entire G-rich region. A region from –289 bp to –249 bp of the cutinase promoter was found to contain a silencer. This 59 bp putative "silencer" element (–295 to –237), when placed in either orientation, exhibited a true silencer-like activity. A GC-rich palindrome that mediates inducible expression of the cutinase promoter was identified. Deletion analysis suggested that the elements required for inducible expression of CAT reside between –209 and –141. Results of PCR-directed mutagenesis within this region showed that inducible expression required –181 to –167 of the promoter. Two overlapping palindromic sequences were found in this region (Fig. 2). CAT activities of transformants harboring PCR-directed mutation constructs showed that palindrome 2 is the one that is essential for inducibility.

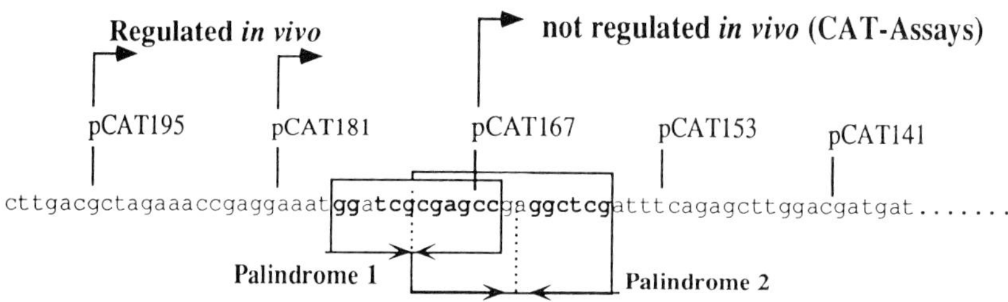

Fig. 2. The deletion constructs used to define the –181 to –167 region essential for inducible expression and the palindromes contained in this region.

A protein designated Cutinase Transcription Factor 1 (CTF1) was found to bind *in vitro* to the GC-rich palindrome involved in cutinase gene induction. That palindrome 2 was indeed the binding site of the nuclear protein(s) was demonstrated by gel shift assays with the derivatives of the –225 to –118 probe containing the same mutations as those used for the transformation studies indicated above.

Methylation interference assays revealed the symmetrically-distributed G residues in palindrome 2 involved in binding of CTF1 (Fig. 3). UV-cross linking of ^{32}P-labeled probe with CTF1 followed by DNAse treatment and SDS-PAGE showed a major band at 49 kDa (monomer) and a minor one at 91 kDa (probably dimer) (249). The fungal protein factor required to get maximal stimulation of cutinase transcription by the cutin monomer showed a native molecular size of 100 kDa and the phosphorylated protein obtained by incubation of the nuclear preparation with the cutin monomer and protein factor showed a monomer size of 50 kDa (390). All results together suggest that the monomeric CTF1 is 50 kDa and it binds as a dimer to activate cutinase gene transcription. CTF1 binding could be demonstrated only when nuclear preparations from glucose-depleted cultures were used, suggesting that the factor may be activated by phosphorylation to bind palindrome 2 only upon glucose depletion.

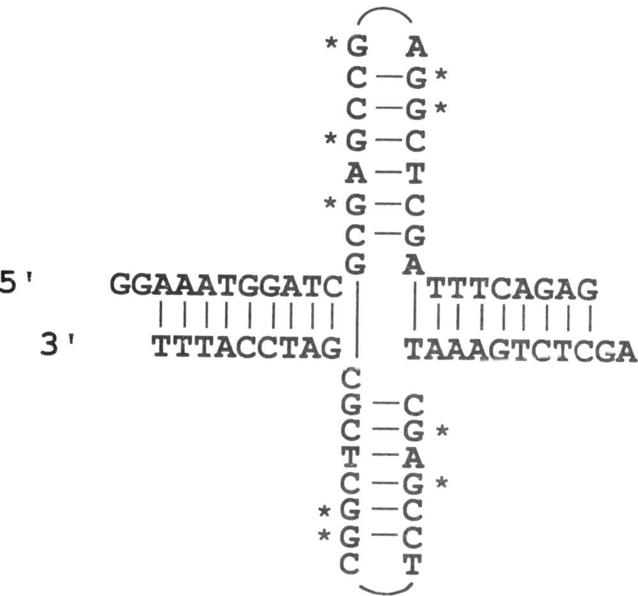

Fig. 3. Methylation interference assay for CTF1-binding region.

A second DNA-binding protein, CTF2, was identified by gel shift assays using a fragment containing the first 116 bp of the promoter. Methylation interference experiments revealed the G residues that are in close contact to this DNA-binding protein, demonstrating that CTF2 binding is between the transcription initiation site and TATA box (Fig. 4).

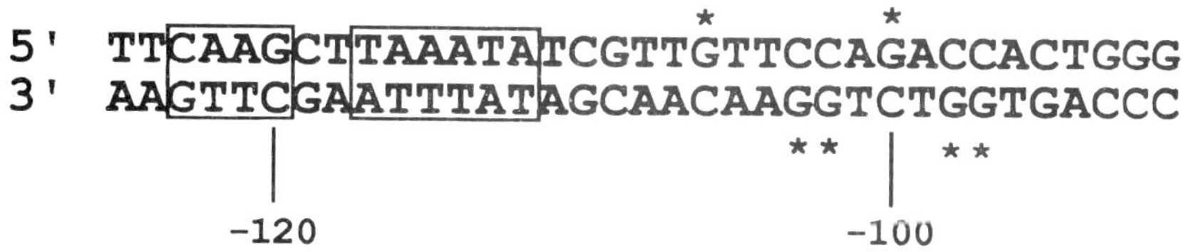

Fig. 4. Methylation interference assay for CTF2-binding region.

Cutin monomers are required for the phosphorylation of transcription factor(s) that binds to the cutinase gene promoter only when phosphorylated

When a nuclear preparation from F. *solani* f. sp. *pisi* was incubated with a fungal protein factor and the unique cutin monomers, cutinase gene transcription was induced in vitro. This transcription activation required a 30 min preincubation of the nuclear preparation with the protein factor and cutin monomer (389). We found that a 50 kDa protein is phosphorylated during this period only if the cutin monomer and fungal protein factor are present (21). This phosphorylation was found to be inhibited by protein kinase inhibitors such as H7 and genestein, a more selective inhibitor of tyrosine protein kinase. In fact, phosphotyrosine in the 50 kDa protein was detected by western blots using antiphosphotyrosine antibodies; formation of phosphotyrosine required the presence of cutin monomer and the fungal protein factor. That this phosphorylation was required for binding to the promoter was demonstrated by the observation that pretreatment of the fungal protein with immobilized phosphatase prevented binding to the promoter (21). Cutinase transcription activation by cutin monomer was inhibited by the presence of kinase inhibitors H7, genestein or antiphosphotyrosine antibodies during the 30 min preincubation period, whereas addition of inhibitors thereafter had little effect. Genestein drastically reduced inducibility of the cutinase gene in protoplasts but did not affect ubiquitin gene expression. Thus, the phosphorylation appears essential for cutinase transcription activation.

Possible role of cyclic AMP in catabolite repression of the cutinase gene

As glucose was depleted, the cAMP level increased about 3-fold from 3.18±1.62 pmol/mg protein to 10.31±2.94 pmol/mg protein in F. *solani* f. sp. *pisi* (Kämper and Kolattukudy, unpublished). Addition of dibutyryl cAMP to protoplasts from non-glucose depleted cultures allowed cutin monomer-dependent induction of the cutinase gene. Thus cAMP may be involved in overcoming catabolite repression. Inhibitors of Ca^{+2}/calmodulin system such as nifedipine, nicardipine, and chlorpromazine inhibited expression of the cutinase gene but not the ubiquitin gene in protoplasts; Ca^{+2} augmented cutinase induction. Although such results suggest Ca^{+2}/calmodulin involvement in cutinase gene transcription, definitive conclusions must await further tests.

When the fungal protein factor required for cutinase gene transcription activation in the isolated nuclear preparation was fractionated by gel filtration, phosphorylation of a protein that depended on cutin monomer and cAMP was detected. The fractions that catalyzed this reaction coincided with the fractions that activated cutinase gene transcription and cross–reacted with antibodies prepared against ATF4, a mammalian transcription factor that binds a cAMP responsive element. Screening of an expression library from induced *F. solani* f. sp. *pisi* cultures with this antibody yielded one clone, fCREB. The protein deduced from fCREB contains a leucine zipper domain that matched with leucine zipper domains found in a variety of transcription factors (Kämper and Kolattukudy, unpublished).

PECTATE LYASE GENES

Penetration of fungal pathogens into plants involves degradation of pectinaceous layers. Even pathogens that penetrate the host through stomata or wounds have to degrade pectinaceous layers to successfully establish infection. Pectin-degrading enzymes are known to be produced by fungi (265). We isolated three different pectin degrading enzymes from F. *solani* f. sp. *pisi*, two pectinases (one exo and one endo) and one pectin lyase (272). When spores were placed on pea stem segments with the lyase antibodies infection was prevented (125). In such bioassays control infection drops resting on pea stem contained preimmune serum while the test droplets contained the antiserum. The only difference between the two was the presence of the specific antilyase IgG. Thus, the lyase appeared to be essential for infection. Similar tests with antibodies prepared against the two hydrolases did not show measurable levels of protection of the host against infection. These results do not necessarily rule out the possibility that the hydrolases play a role in pathogenesis because it is possible that antibodies could not reach the deeper cell wall areas where the hydrolases might play a role.

Both cDNA and the gene for the previously purified pectate lyase from *F. solani* f. sp. *pisi* were cloned and sequenced (190). This lyase sequence showed little homology to those of other pectolytic enzymes. RNA blot analysis showed a single band of mRNA at about 1 kb. Pectate lyase A (*pelA*) gene expression was shown to be induced by pectin and repressed by glucose. Southern blot analysis under low stringency conditions showed several hybridization bands. Cloning and sequencing of such DNA fragments showed the presence of three additional *pel* genes designated *pelB, pelC* and *pelD* with 63%, 46% and 43% overall peptide homology, respectively, to *pelA*; *pelA* and *B* but not *pelC* and *D* contained N-terminal signal sequence (Guo, González-Candelás and Kolattukudy, unpublished). RNA blot analysis and PCR approaches using gene specific primers detected only *pelA* transcripts when *F. solani* f. sp. *pisi* was grown on pectin. Regulation of expression of the different members of the family of lyase genes and their role in pathogenesis are yet to be elucidated.

SUBERIZATION IN PLANT-FUNGUS INTERACTIONS

Suberization of cell walls is a process that plants use to erect barriers to respond to a variety of developmentally- and environmentally-imposed needs (262, 266). The chalazal region of seed coats and Casparian bands are examples of the former (262); wound healing and fungal attack are examples of the latter (412). Suberization involves deposition of insoluble polymeric materials and soluble waxes on the cell wall (266). From a series of studies on the structure, composition and biosynthesis of suberized walls from a variety of anatomical regions, a working model for suberized walls was proposed (261, 262). According to this model (Fig. 5), the polymeric part of suberin consists of aromatic domains attached to the cell walls and aliphatic domains attached to the aromatic domains; soluble wax is associated with the suberized wall to make the wall resistant to diffusion. Because of the complexity of the isolated suberized cell walls, it has been difficult to get direct evidence concerning the structure.

Recently CP/MAS ^{13}C NMR has been used to get some structural information (177, 268, 270). On the whole, these results support the proposed model. Thus, carbohydrates that constitute about one-half of the isolated suberized walls (262) form a separate structural domain as expected from attachment of a new layer to the cell wall (177). Aliphatic domains were found to be much less abundant than aromatic domains (177, 262). More direct structural information is necessary to test and refine the model.

Biosynthesis of suberin involves synthesis of the monomers of the aromatic and aliphatic domains. The aromatic monomer synthesis that starts with phenylalanine ammonia lyase has been extensively studied (196, 199) and similar mechanisms are most probably involved in suberizing cells. Except for the unique components such as the dicarboxylic acids which are produced by a specific dehydrogenase (5), aliphatic components are probably synthesized using mechanisms similar to those involved in cutin synthesis. The most unique aspect involves formation of the polymer from monomers. A variety of lines of evidence suggested that the polymerization of the aromatic components involves highly anionic peroxidase(s) (262, 266). It was suggested that conjugates of aliphatic components with aromatics, found in periderm (1, 50), could be substrates for such peroxidases (262). However, this postulate has not been tested directly. The anionic peroxidase postulated to be involved in suberization was purified from wound-healing potato periderm. Immunological localization of this enzyme showed that it is present only at the site of suberization and only during suberization (159). The cDNA and gene for this peroxidase were cloned and time-course of appearance of the transcripts correlated with that of suberization in wound-healing potato tuber and tomato fruits (413, 414).

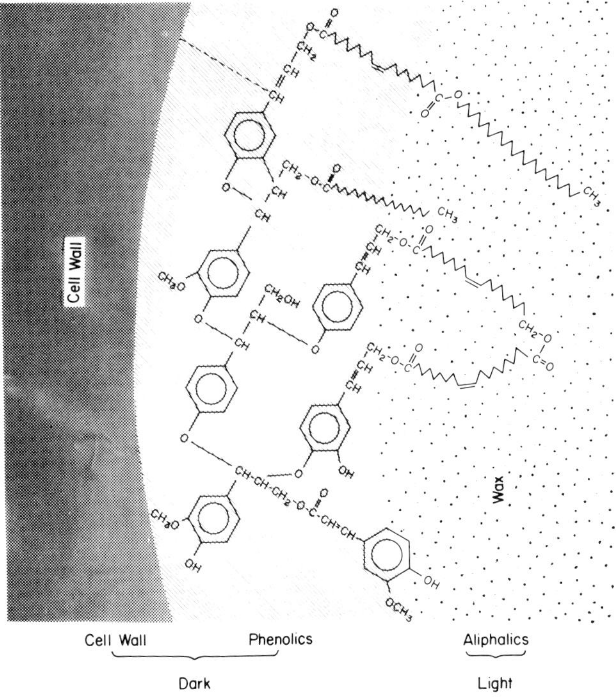

Fig. 5. A working model for the structure of suberin. Examples of possible linkages are shown.

More than a decade ago we proposed that reinforcement of cell wall by suberization might be a possible mechanism to limit ingress of pathogenic fungi into the host (260). We recently provided evidence for this hypothesis. Vascular coatings formed by tomato plants in response to *Verticillium alboatrum* or abscisic acid were shown to be suberin by histochemical and direct GC/MS chemical analysis (412). Of two isolines of tomato *(Lycopersicon esculentum* L. cultivar Craigella) that differ in susceptibility to the pathogen, only the resistant one showed suberization when attacked by the fungus. We cloned and sequenced the two tomato anionic peroxidase genes *(tap1* and *tap2)* and demonstrated their induction by abscisic acid. RNA blots for suberization-associated anionic peroxidase mRNA showed induction of this peroxidase gene and tissue blot analysis indicated that induction of the mRNA was localized in the responding vascular bundles. Differential expression of peroxidase in the two near-isogenic tomato lines could also be demonstrated with cell cultures. Cell cultures from the resistant line, when treated with low levels (ng/ml) of fungal elicitor, generated the peroxidase mRNA within minutes, whereas cells from the susceptible line hardly responded (330). In collaboration with Dr. P. Balint-Kurti (Sainsbury Laboratory, John Innes Institute), the *tap* genes were localized on chromosome 2 in the tomato genome and it appears that *tap1* and *tap2* might be *prx2* and *prx3* that had been localized 0.14 cM apart.

The availability of the cloned anionic peroxidase genes allowed us to use transgenic plants to test for the roles of this peroxidase. In transgenic tobacco plants containing *tap1* the introduced gene was transcribed in a wound-inducible manner. In transgenic tobacco plants containing *tap*1/GUS and *tap*2/GUS fusion constructs, both *tap*1 and *tap*2 were induced at the site of wounding and fungal attack. Evidence with transgenic plants indicates that the peroxidase gene begins expression soon after wounding, suggesting that the anionic peroxidase could reinforce the walls by causing cross-link formation between wall proteins and wall-bound phenolics early (329), by deposition of the phenolics generated by the first wave of synthesis that happens within 24 hr after wounding (196, 199), and finally by the deposition of the aromatic domains of suberin.

A series of nested 5' promoter deletions in the *tap*1/GUS fusion gene construct were tested in tobacco protoplasts (transient expression) and in transgenic tobacco plants. Both approaches showed that induction of *tap*1 required 358 bp 5'-flanking region of the gene whereas constitutive expression required only 202 bp (331). Constitutive GUS expression was found in the epidermis, trichomes, leaf traces, pistils, ovaries and anthers in transgenic plants. The developmentally-regulated tissue specific expression of GUS was found with all constructs with the –202 bp and larger promoters whereas wound and pathogen induction required –358 bp or larger promoters. These results suggest that the *tap*1 gene, heretofore thought to be expressed only upon wounding or pathogen attack, plays a role in normal developmental processes of the plant and this gene acquired additional 5'-flanking promoter for the purpose of responding to wounding and fungal attack.

Over-expression of *tap*1 gene with double CMV 35S promoter resulted in a 10- to 400-fold increase in peroxidase activity in the transgenic tobacco plants which appeared normal in all respects showing no wilting at any time (Sherf and Kolattukudy, unpublished). Highest levels of expression were observed in the vascular region and epidermis. Both 35S-driven *tap*1 expression and *tap*1 expression due solely to its own promoter conferred resistance to blue mold in transgenic tobacco; the constitutive expression gave near complete protection. Constitutive peroxidase expression also provided some protection to potato plants. These peroxidase constructs have been introduced into mustard and rice, and are being introduced into peanut and cotton among others for evaluation of the effects.

CONCLUSION

The host defensive barriers need to be broken for the fungus to invade the host. Since disease in most cases involves damage to the host in multiple ways no single component may be identified as the single virulence factor. Those fungi that evolved with successful strategies have multiple genes and mechanisms to degrade the barriers. Redundant sets of genes with different regulatory triggers to produce extracellular enzymes appear to be a common strategy. In response to such attack, the host has evolved complex mechanisms to limit the ingress of the pathogens by reinforcing the defensive barriers by a series of complex processes that initiates immediately and proceeds until the damage site is sealed off from the host by suberization. Understanding of mutual triggering of gene expression and consequent process could lead to novel ways to intervene in this process to help protect plants.

ACKNOWLEDGMENTS

We thank Dr. W. Schäfer for kindly providing the cutinase gene-disrupted transformant and Paula Pack for her assistance in preparing this manuscript. This work was supported by NSF Grant DCB 8819008.

FUNGAL CHITINASES: THEIR PROPERTIES AND ROLES IN MORPHOGENESIS, MYCOPARASITISM AND CONTROL OF PATHOGENIC FUNGI

Mohan S. MANOCHA[1] and R. BALASUBRAMANIAN[2]

[1]Department of Biological Sciences, Brock University, St. Catharines, Ontario, Canada L2S 3A1
[2]Centre of Advanced Study in Botany, University of Madras, Madras 600 025, India

Chitinase hydrolyzes chitin, the insoluble unbranched homopolymer of N-acetylglucosamine (GlcNAc) in a ß-1,4 linkage. The degradation of chitin occurs in two steps. In the first step, chitin microfibrils are cleaved by either endo- or exochitinase, or both. Endochitinase cleaves chitin randomly, resulting in soluble low molecular weight GlcNAc oligomers such as chitotetraose and chitotriose with the dimer, chitobiose, being predominant. Exochitinase, on the other hand, catalyzes the progressive release of chitobiose in a stepwise fashion and no monomer or oligomers are formed. In the second step, chitobiose is hydrolyzed to GlcNAc monomers by ß-1,4 N-acetylglucosaminidase. ß-1,4 N-acetylglucosaminidase can also hydrolyze chitotriose and chitotetraose at decreasing rates as the number of GlcNAc residues increases.

Chitin is the second most common biodegradable polymer in nature. It has a very wide distribution as a tough exoskeletal material in insects and crustaceans (242). Chitin is almost ubiquitous in fungi, with only few exceptions. It is an integral component of most fungal cell walls and the sole component of the primary septa in yeast. Chitinase has been implicated in a number of morphological and physiological processes that occur during the life cycle of a fungus. Chitinase also seems to play a role in the early events of host-parasite interaction in biotrophic and necrotrophic mycoparasitism. Chitinase involvement in similar events between entomopathogenic fungi or vesicular arbuscular mycorrhizal fungi and their hosts has also been reported (146, 468).

This review provides information on recent developments in research on fungal chitinases and deals primarily with their role in mycoparasitism and biocontrol of plant pathogenic fungi.

GENERAL PROPERTIES OF CHITINASES

Some properties of chitinases from various fungi are summarized in Table 1. Chitinases are active at slightly acidic pH, have high temperature optima and a high degree of stability, which may be due to glycosylation. Their molecular masses differ considerably in various fungi. Chitinases are inhibited by copper and mercury salts and competitively by chitobionolactone oxime (398) and allosamidin whose inhibitory potency is pH dependent (325).

Chitinase activity is stimulated by partial proteolysis of microsomal fractions with commercial proteases such as trypsin or by partially purified proteases from the fungus itself (235, 305). Humphreys and Gooday (234) reported inactivation of microsomal chitinase from *Mucor mucedo* after treatment with commercial phospholipases, thus suggesting the requirement of a phospholipid environment for enzyme activity.

LOCALIZATION OF CHITINASE

Chitinases as soluble proteins are found in different cellular fractions, i.e. in vacuoles, membrane-bound or cell wall-bound, of various fungi (137, 235, 237). Information on the spatial localization of chitinase in the fungal cell is necessary to elucidate its role in various morphogenetic events. Sahai et al. (428) used a polyclonal antibody prepared against a purified microsomal chitinase of *Choanephora cucurbitarum* and the technique of immunofluorescence and reported localization of chitinase at various developmental stages of five zygomycetous fungi. Their results indicated an involvement of chitinase in spore swelling, germination, sporangium development, and response to mechanical injury. Chitinase was not detected at the apices of the hyphae of all the five fungi tested. Binding with fluorescein isothiocyanate (FITC) labelled wheat germ agglutinin (WGA) revealed patterns of fluorescence similar to that of chitinase localization, but differed by showing fluorescence and therefore N-acetyl-glucosamine (GlcNAc) oligomers at the hyphal apices. Differences in binding ability of antichitinase and the lectin WGA suggests that the latter is not a suitable indicator for indirect localization of chitinase as reported previously (109).

Subcellular localization of chitinase using the technique of colloidal-gold immuno-cytochemistry in the two mucoraceous fungi, *Mortierella pusilla* and *Phascolomyces articulosus*, susceptible and resistant hosts, respectively, to the mycoparasite, *Piptocephalis*

Table 1. General properties of various fungal chitinases.

Organism	Enzyme location	pH optima	Temperature optima	Molecular mass (kDa)	Inhibitory agents[c]	References
Choanephora cucurbitarum	IC	6.0	40°C*	55 & 69.5	Mn^{2+}, KCN	23
	MS	4.5–6.0	40°C*	66	Mn^{2+}	
Mucor mucedo	IC	5.65	–	–		235
	MS	5.55	–	–	PMSF	
Mucor rouxii	IC	6.5	30°C	30.7		398
Phascolomyces articulosus	IC	–	40°C*	53 & 69.5	Mn^{2+}, KCN	23
	MS	–	40°C*	66	Mn^{2+}, EDTA	23
Neurospora crassa	IC	6.7	–	20.6		529
Saccharomyces cerevisiae	IC	2.5	–	130	–	283
Aspergillus nidulans	EC	5.0	–	27	Ca^{2+}, Ag^{2+}	402
Candida albicans	IC	6.5	45°C*	70	allosamidin	27, 137
	MS	8	45°C*	–	allosamidin	
Aphanocladium album	EC	4.0	50°C*	39[a], 20[b]	Fe^{2+}	282
Metarhizium anisopliae	EC	5.3	38–45°C*	33 & 66[b]	–	468
Trichoderma harzianum (strain 39.1)	EC	4.5	40°C	40[a], 36	Zn^{2+}	496
Verticillium albo-atrum	IC	5.1	45°C	64	KCN	371
Myrothecium verrucaria	EC	4.0–6.5	30–50°C	40[a]	$ZnS0_4$,$CaCl_2$	150
Acremonium obclavatum	EC	3.0–4.0	40–50°C	45[a]	$MnCl_2$,$ZnS0_4$	151

EC, extracellular; IC, intracellular; MS, microsomal; *stable at low temperature; [a] by SDS-PAGE; [b]by gel filtration; [c] all chitinases are inhibited by copper & mercury salts, additional inhibitors are given in the table. Modified from Sahai and Manocha (1993). FEMS Microbiology Reviews 11:317–338.

virginiana, revealed that in *M. pusilla* the chitinase is mainly vacuolar, whereas in *P. articulosus* it is also localized in the cell wall (Manocha, unpublished results). Whether these differences in spatial localization of chitinase can be correlated with host cell wall characteristics which may account for the extent and nature of parasitism by *P. virginiana* is under investigation in our laboratory.

REGULATION OF CHITINASE

Chitinase production in fungi seems to be regulated by a repressor-inducer system wherein chitin or its oligomers serve as inducers. High levels of chitinase were found in *Metarhizium anisopliae* cultures supplied with chitin but not pectin or cellulose (467). Of the inducers tried, GlcNAc was the most effective. Glucosamine could also induce chitinase since chitin from natural sources appears to be partially deacetylated while chitosan contains approximately 10-20% acetylated residues. In the entomopathogenic fungus *Beauveria bassiana*, chitinase is an inducible enzyme and repressed by glucose (461). Induction of chitinase by GlcNAc and repression by glucose also occurs in a chitinase overproduction mutant of *Aphanocladium album,* strain E3 (502). Examination of the regulation of chitinase gene expression in *A. album* by Northern blot analysis revealed induction of chitinase 1 mRNA grown in the presence of chitin and repression of gene expression by glucose (61). Studies on chitinase regulation in *Trichoderma harzianum* (strain 39.1) by Ulhoa and Peberdy (496) registered high activity of chitinase only in cultures supplied with chitin but not with cellulose, chitosan or chitobiose. Unlike the induction of chitinase by GlcNAc in the above mentioned fungi, the *T. harzianum* enzyme synthesis was repressed by both GlcNAc and glucose. Thus, repression of chitinase synthesis by glucose in *A. album*, *B. bassiana* and *M. anisopliae* by glucose, and GlcNAc in *T. harzianum* 39.1 suggests the involvement of catabolite repression in the regulation of chitinase synthesis. Regulation of chitinolytic enzyme production by catabolite repression in *M. anisopliae* is a case in point (467). Evidently, fungal chitinases, like other polysaccharidases provided with complex exogenous substrates, are inducible and subject to catabolite repression.

MOLECULAR BIOLOGY OF CHITINASES

Kuranda and Robbins (283) sequenced the endochitinase gene of *Saccharomyces cerevisiae*. Amino acid sequence analysis indicated that there are four domains, namely a signal sequence, a catalytic domain, a serine/threonine rich region and a chitin binding domain at the carboxy terminal. Interestingly, the catalytic domain was shown to be homologous to a pathogenesis related cucumber chitinase. There are also small regions of this domain conserved with the cucumber chitinase, bacterial chitinase, endoglycosidase H, and a mammalian lysosomal chitinase. All of them are able to cleave the ß-1,4 glycosidic bond between adjacent GlcNAc residues.

Elucidation of the chitin binding domain is significant in view of the fact that a number of polysaccharidases have noncatalytic high affinity binding domains for their respective substrates. Thus, the chitin binding domain is not only required to localize the enzyme on the yeast cell wall but is also involved in cell separation (90, 283). Recently, the deduced amino acid sequences of *Rhizopus oligosporus* chitinase genes have been shown to be similar to that of the published sequence of *S. cerevisiae* chitinase. *R. oligosporus* chitinase, however, had an additional C-terminal domain (526). Analyses of the N-terminal and C-terminal amino acid sequences of both chitinases and their comparison with the amino acid sequences deduced from the nucleotide sequences revealed post translational processing not only at the N-terminal signal sequences but also at the C-terminal domains (Fig. 1).

Blaiseau et al. (60) have sequenced a chitinase gene from *Aphanocladium album*. Antibodies were raised against chitinase 1 (chi 1) from the chitinase overproducing mutant of *A. album* and selected a suitable cDNA clone using standard techniques. Transformation of two strains of *Fusarium oxysporum* f. sp. *melonis* with the chitinase genomic fragment resulted in enhanced extracellular chitinase activity over the control but did not inhibit growth. Blaiseau and Lafay (61) have determined the complete sequence of the chromosomal and cDNA copies of the gene coding for chi 1. Data on amino acid sequence of the chi 1 gene show similarities with bacterial chitinases from *Serratia marcescens* and *Bacillus circulans*. Compared to other chitinases, *A. album* chi 1 has only two short similarity regions which are also found in bacterial, yeast and some plant chitinases. Knowledge of isolation and sequencing of genes encoding chitinases will not only reveal evolutionary aspects of fungal chitinases but also provide clues to the understanding of evolution of plant chitinases. A comparison of complete amino acid sequence of fungal chitinase with that of a plant chitinase would be worth further study in view of the large body of evidence pointing to plant chitinases as potent inhibitors of fungal growth.

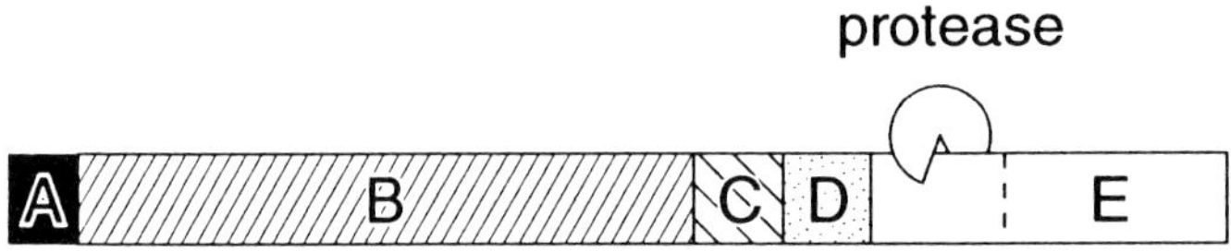

Fig. 1. Schematic representation of *R. oligosporus* chitinases redrawn from (526), with permission. **A**, signal sequence; **B**, catalytic domain; **C**, Ser/Thr-rich domain; **D**, chitin-binding domain; **E**, the C-terminal domain with two regions and protease processing site.

ROLE OF CHITINASE IN MORPHOGENESIS

The unitary model of cell wall growth of Bartnicki-Garcia (28) envisages the role played by lytic enzymes in maintaining a balance between wall synthesis and wall lysis during hyphal apical growth, maintaining the apex in a plastic state and permitting insertion of nascent chitin into the wall.

Proof for the association of chitinases and chitin synthases comes from parallel behaviour of the two activities during spore germination in *Mucor mucedo* (191), during exponential growth in *M. rouxii* (398) and *Candida albicans* (27), and from the finding of a membrane-bound chitinase activity in the same cell fraction as chitin synthase in *M. mucedo* (234, 235). With the aid of immunofluorescence techniques, Sahai et al. (428) have provided direct evidence for the presence of chitinase during spore swelling, germination, sporangium formation and response to mechanical injury in *Choanephora cucurbitarum* and four other Zygomycetes fungi. Failure to localize chitinase at the hyphal apices suggests its possible lack of involvement in apical growth.

In *Saccharomyces cerevisiae* partial activity is found in the periplasmic space (151) and chitinase is secreted into the medium by protoplasts incubated in growth medium. The enzyme activity is highest during logarithmic phase of growth, thereby suggesting a role during septum formation and cell separation. Evidence for its role in cell separation comes from disruption of a cloned *S. cerevisiae* chitinase gene, which resulted in a defect in cell separation (283). The chitinase requirement for lysis is confirmed by the protecting ability of demethyl allosamidin against lysis and the disruption of the chitinase gene in ChS1 cells which eliminates lysis (90).

The process of autolysis of mature fruiting bodies of *Coprinus lagopus* is accomplished by the action of chitinases which are formed shortly before spore release begins (237). Demonstration of lysosomal chitinases was based on sedimentability of hydrolases. Thus, chitinase activity was localized intracellularly in vacuoles together with other hydrolytic enzymes. Chitinases had apparently no function in intracellular digestion since they were synthesized shortly before autolysis in gills (237). This hydrolase is passively released into the wall upon cessation of metabolic activity in senescing cells.

Mahadevan and Mahadkar (301) and Rosenberger (416) described autolytic enzymes including chitinases that are bound to sub-apical walls of *Neurospora crassa* and *Aspergillus nidulans* and suggested that chitinases are associated with hyphal branching rather than autolytic wall turnover.

Fungal chitinases have so far been implicated in apical growth, spore swelling and germination, liberation of spores, cell separation and budding in the yeast system. It seems very likely that chitinases of fungi may also play a role in hyphal fusion. With the development of immunological techniques at the electron microscope level, using Protein A colloidal gold labelling, it will be possible to ascribe a function for chitinases in hyphal anastomosis. To this end, several laboratories have developed antichitinase antibodies against purified fungal chitinases which will enable them to evaluate the role of chitinases in morphogenesis of fungi. Serological comparison of fungal chitinases of different groups may reveal affinities (relationships) among them and can also be used as a tool in the classification (serotaxonomy) of chitin containing fungi.

ROLE OF CHITINASES IN MYCOPARASITISM

Necrotrophic Mycoparasitism

The antagonistic ability of *T. harzianum* Rifai against several soilborne plant pathogens and the involvement of lytic enzymes in the interfungal antagonism was investigated in several laboratories. The enzymatic basis of the interaction between *T. harzianum* and plant pathogens was found to be due to the activity of chitinases and glucanases produced by *T. harzianum* on isolated mycelium of

Sclerotium rolfsii, Rhizoctonia solani and *Phythium aphanidermatum* (150). Elad et al. (150) concluded that production of lytic enzymes can be used as a tool for screening highly parasitic isolates of *Trichoderma* in soil. Based on ultrastructural observations and cytochemical localization of N-acetylglucosamine residues, Cherif and Benhamou (109) provided indirect evidence that chitinase production may have significance in the mycoparasitism of *Trichoderma* on *Fusarium oxysporum* f. sp. *radicis-lycopersici.* However, mechanism other than lytic enzyme production may be involved in the interaction between *Trichoderma* sp and *Fusarium oxysporum*. According to Ordentlich et al. (358) specific recognition, production of antibiotics, and competition for nutrients may all play a role in the biological control of Fusaria. In any case, a chitinase of *T. harzianum* has also been shown to produce lytic halos on purified cell walls of another phytopathogen, *Botrytis cinerea* (131). An extracellular chitinase produced by *Myrothecium verrucaria* inhibits germination and germ tube growth of the groundnut rust fungus *Puccinia arachidis* (Balasubramanian, unpublished results). Similarly, another mycoparasite of *P. arachidis*, *Acremonium obclavatum*, produces an extracellular chitinase *in vitro* which inhibits germination of uredospores of the peanut rust (197). These observations show the possible role played by chitinases in necrotrophic mycoparasitism. Conclusive proof, however, must probably await the results of *in situ* experiments using chitinase-deficient mutants.

Biotrophic Mycoparasitism

Penetration of fungal hosts by the biotrophic mycoparasite *Piptocephalis virginiana* (304) occurs by both mechanical and enzymatic mechanisms. Light and scanning electron microscopic studies have shown inpushing of the susceptible host cell wall and enzymatic erosion of the resistant host cell wall by the advancing infection hyphae (307). Culture filtrates of *P. virginiana* contained only negligible levels of chitinases and chitosanase thus indicating strict regulatory control of these lytic enzymes, which is characteristic of a biotrophic mycoparasite (22). Manocha (304) surmised that metabolic shifts favouring chitinase occur in the susceptible host, *C. cucurbitarum* and chitin synthetase in the resistant host, *P. articulosus* when attacked by *P. virginiana*. Enhanced levels of chitinase activity induced in *C. cucurbitarum* may culminate in increased plasticity of the host cell wall and a lack of incorporation of a chitin precursor at the penetration site due to degradation of nascent chitin by chitinase. By contrast, higher levels of chitin synthetase may be present in *P. articulosus* because of deposition of chitin and papilla formation at the penetration sites (304).

Biological Control

Interfungal antagonism, responsible for biological control, may involve antibiosis, competition, or mycoparasitism (110). One of the attributes of mycoparasitism is the production of lytic enzymes for the degradation of cell walls of the parasitized fungi (458). Reports dealing with biological control of soil-borne pathogens suggest that chitinolytic enzymes play a major role in plant protection (357). Chet et al. (111) used genetic engineering techniques to prove the involvement of chitinase in the control of *Sclerotium rolfsii* by *Serratia marcescens*. A gene, Chi A, encoding for the major chitinase produced by *S. marcescens*, was cloned and sequenced. *Escherichia coli* cells harboring plasmids ligated with the Chi A gene were found to synthesize chitinase following prolonged incubation (480, 481). Subsequently, an efficient expression of chitinase in *E. coli* was constructed by Shapira et al. (440). A DNA fragment carrying the Chi A gene from *Serratia marcescens* was subcloned into the plasmid pBR322.

Evidence for the role of chitinase in plant protection against fungal infection is multiple. The protein isolated from *E. coli* carrying the plasmid pCH1A lacking the PL promoter contained low levels of chitinase activity and was rather ineffective in protecting bean seedlings. Secondly, partially purified chitinase produced by the cloned gene induced rapid and extensive bursting of the hyphal tips of *S. rolfsii*. Finally, plants irrigated daily with the chitinase preparation were significantly protected from *Sclerotium rolfsii* infection throughout (111).

Another approach has been the successful introduction of the Chi A gene into tobacco plants from the soil bacterium *Serratia marcescens*. Chi A-tobacco transgenic plants grew nearly as fast as uninfected plants and showed significantly better protection against *Rhizoctonia* infection than the control plants (238). Although there have been reports on the overexpression of plant chitinases (85, 295, 338) or of a chitinase derived from the commercially available *Serratia marcescens* strain QMB 1466 (299, 486) in transgenic tobacco plants, the first report describing the protection of tobacco plants against fungi is the work of Broglie et al. (85).

Introduction of chitinase from a fungal biocontrol agent such as *Trichoderma harzianum* or *Gliocladium* sp. into higher plants to confer protection against fungal infection is an interesting possibility in the years to come. The time is ripe for such a development in view of the potential for biocontrol by *Trichoderma* and *Gliocladium* against soil borne disease. The expression of chitinases from an unrelated species, in transgenic plants, will lead to "new" chitinases which cannot be overcome easily by the invading fungus.

ACKNOWLEDGMENTS

MSM acknowledges financial assistance from the Natural Sciences and Engineering Research Council of Canada in the form of an operating grant. Thanks are also due to our graduate students for their help in research.

PLANT CELL WALL STRUCTURAL PROTEINS: REGULATED EXPRESSION AND ROLES IN FUNGAL INFECTION

Jinsong SHENG and Allan M. SHOWALTER

Department of Environmental and Plant Biology, Molecular and Cellular Biology Program, Ohio University, Athens, OH 45701, USA

Primary cell walls are thin and strong, properties conferred by the arrangement of cellulose fibers embedded in a matrix of polysaccharides and proteins. This matrix contains mainly two polysaccharides, hemicellulose and pectin. Several wall proteins are also present and include extensins, proline-rich proteins (PRPs), glycine-rich proteins (GRPs), solanaceous lectins, and arabinogalactan proteins (AGPs) (reviewed in 451); their biochemical properties are summarized in Table 1. Based on their abundance, their extremely biased amino acid compositions, and the presence of repeating amino acid sequence motifs, all five protein classes are considered to be structural proteins. cDNA and gene clones are characterized for all these protein classes except the solanaceous lectins and AGPs.

Primary walls are dynamic entities, with wall polymers changing in types, amounts, and associations throughout development and in response to a variety of stress conditions, including those instigated by fungal infection. Wall architecture is thus altered presumably to accommodate growth or, in the case of fungal infection, to erect a defensive barrier. Here we will examine the structure and expression of various wall proteins in response to fungal infection and hypothesize on their roles and the signal transduction scheme(s) by which they are affected during fungal infection. To date, the extensins and PRPs are the only two cell wall protein classes which have undergone serious examination with respect to fungal infection; thus our discussion is largely restricted to these two protein classes.

SOME EXTENSIN mRNAs ARE INDUCED BY FUNGAL INFECTION AND ELICITOR TREATMENT

Accumulation of extensins was first observed in cell walls of melon (*Cucurbita*) seedlings inoculated with *Colletotrichum lagenarium*. The level of extensins reached 5 to 15% of the wall mass upon infection, compared to 0.5% in healthy tissue. Accumulation of extensins was consistent with the subsequent observation of

Table 1. Five major classes of structural plant cell wall proteins and some distinguishing properties.[a]

Protein Class	% Protein	% Sugar	Abundant amino acids[b]	Amino acid motifs[b]
Extensins (dicot)	~45	~55	O,S,K,Y,V,H	SOOOOSOSOOOOYYYK SOOOOK SOOOOTOVYK SOOOOVYKYK
"Extensins" (monocot)	~70	~30	O,T,S,P,K	TPKPTOOTYTOSOKPO
				ATKPP
GRPs (dicot)	~100	~0	G	GX
GRPs (monocot)	~100	~0	G	GG and GGY
PRPs	80-100	0-20	O,P,V,Y,K	PPVYK and PPVEK
PRPs (nodulins)	?	?	O?,P,E,Y,H,K	PPHEK and PPPEYQ
Solanaceous lectins	~55	~45	O,S,G,C	$S(O)_{2\text{-}6}$[c]
AGPs	2-10	90-98	O,S,A,T,G	AO

[a] Modified from Showalter (451). AGP: arabinogalactan proteins; GRP: glycine-rich proteins; PRP: proline-rich proteins. [b] O=Hydroxyproline. [c] M. Kieliszewski, personal communication.

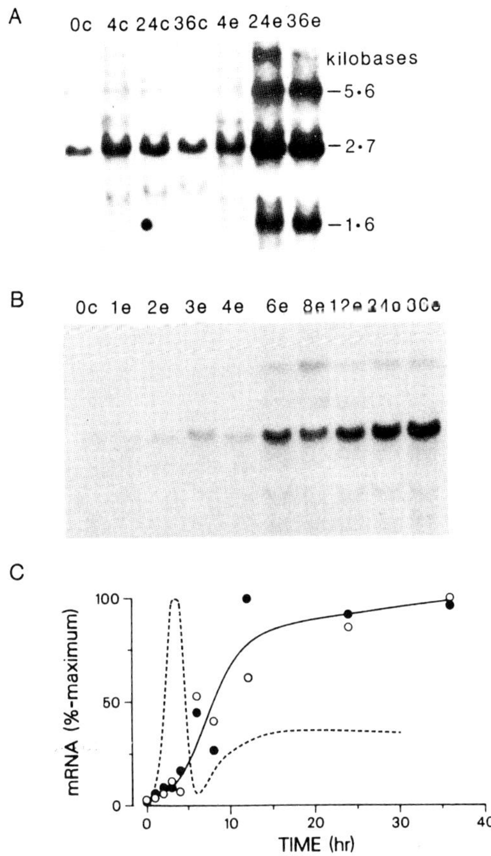

Fig. 1. Accumulation of extensin mRNAs in elicitor-treated bean cells. RNA was blot hybridized with a ^{32}P-labeled tomato extensin genomic probe. **A** and **B**. Total RNA isolated from cells at the times (in h) indicated after elicitor treatment (e) or from equivalent mock-treated unelicited control cells (c). **C**. Kinetics of elicitor-induced accumulation of the 2.7 kb extensin mRNA in total cellular RNA (o) and polysomal RNA fractions (•). Dotted line denotes the kinetics previously observed (425) for the accumulation of mRNA encoding the phytoalexin biosynthetic enzyme chalcone synthase in the same set of RNAs.

increased extensin mRNA levels in the infected cells (312). In melon, bean and tomato plants infected with selected fungal pathogens, highly localized accumulation of extensins was detected in the walls of living cells adjoining dead, hypersensitive cells and also in papillae encasing intracellular fungal hyphae (44, 354).

When bean (*Phaseolus vulgaris*) suspension-cultured cells were treated with elicitor prepared from mycelial walls of *Colletotrichum lindemuthianum* for varying times, marked accumulations of three extensin mRNA species of approximately 5.6, 2.7, and 1.6 kb in size were observed that hybridized to a radiolabelled tomato extensin probe (452) (Fig. 1). The accumulation kinetics of the 2.7 kb extensin mRNA species was distinct from that of chalcone synthase (CHS) mRNA in response to fungal elicitor. Elicitor treatment caused a rapid but transient increase in CHS mRNA, while the extensin mRNAs displayed a less rapid but sustained increase in mRNA levels.

The same extensin mRNAs were also induced by infection of *P. vulgaris*, cv. Kievitsboon Koekoek hypocotyls. In the incompatible interaction (host resistant) with *C. lindemuthianum*, race β, the 2.7 kb and 1.6 kb extensin transcripts accumulated early in the infection. This accumulation was first observed 52 h after inoculation and was most pronounced in tissue at the infection site. Adjacent tissue also responded to the infection but to a lesser extent (Fig. 2A). In the compatible interaction (host susceptible) with *C. lindemuthianum*, race γ, the accumulation of 2.7 kb and 1.6 kb extensin mRNAs occurred, but was substantially delayed in comparison to the incompatible interaction (452) (Fig. 2B).

Corbin et al. (124) characterized three elicitor and infection induced extensin cDNA clones from *P. vulgaris*. These structurally similar extensin cDNAs are differentially regulated in terms of their kinetics of elicitor- and infection-induced mRNA accumulation. A striking feature of these cDNA-encoded extensins is the abundance and clustering of tyrosine residues, almost exclusively in $(Tyr)_3$ blocks occurring within the larger hydroxyproline-rich repeat motifs. Such an organization of tyrosine residues makes the formation of isodityrosine cross-linking between extensins a distinct possibility (451).

For GRPs, there is no information on whether these proteins are regulated in response to fungal infection. However, GRP mRNAs were observed to accumulate in response to infection by tobacco mosaic virus in tobacco and petunia (294, 498).

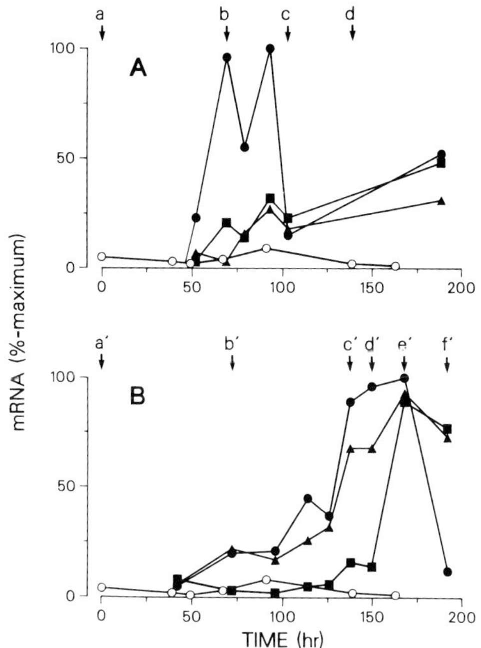

Fig. 2. Kinetics of accumulation of the extensin 2.7 kb mRNA in hypocotyls of bean cv. Kievitsboon Koekoek during race:cultivar-specific interactions with physiological races of *C. lindemuthianum*. **A**. Incompatible interaction with race β. **B**. Compatible interaction with race γ. RNA was isolated from site 1 (●), the direct inoculation site; site 2 (▲), tissue laterally adjacent to the infection site and site 3 (■), tissue directly beneath sites 1 and 2. The accumulation kinetics in equivalent uninfected hypocotyls is indicated by open circles (O). Arrows in **A** denote events in the onset of hypersensitive resistance at site 1: a, spore inoculation; b, onset of hypersensitive flecking in a few sites; c, hypersensitive flecking apparent at most sites; d, very dense brown flecking at all sites. No visible changes occurred in sites 2 and 3 and in control hypocotyls throughout the time course. Arrows in **B** denote events in lesion development at site 1: a', spore inoculation; b', no visible symptoms (cf. incompatible interaction); c', onset of symptom development at a few sites; d', pale to medium brown lesions apparent at most sites; e', onset of water soaking and development of spreading lesions; f', extensive water soaking and spreading of lesions from site 1, some browning at site 2.

SOME EXTENSIN AND PRP mRNAs ARE DOWN-REGULATED BY FUNGAL ELICITOR AND INFECTION

The studies described above have revealed up-regulation of extensin mRNAs and proteins in response to elicitor treatment and infection; however, more recent studies have shown that at least some extensin and PRP mRNAs are down-regulated under these same conditions. For example, a bean extensin mRNA, corresponding to a cDNA named Hyp2.11, was less abundant following elicitor treatment and hypocotyl infection with *C. lindemuthianum* (435). Hyp2.11 encodes an extensin with a low tyrosine content (i.e., 6%) compared to other cDNA-encoded extensins (e.g., Hyp4.1 and Hyp3.6 with 18% and 19% tyrosine, respectively); thus the Hyp2.11 encoded extensin is probably less likely to form isodityrosine cross-links.

A PRP mRNA from *Phaseolus vulgaris*, PvPRP1, is down-regulated by elicitor treatment (443). PvPRP1 mRNA is very abundant in suspension cultured cells, and it is also known to be involved in various developmental processes. Upon elicitor treatment, the PvPRP1 mRNA level is unchanged for 1 h, then decreases dramatically to an exceptionally low level after 10 h (less than 1% of the unelicited, control cells), and remains undetectable up to 24 h after elicitor treatment (Fig. 3). The PvPRP1 protein sequence contains few tyrosine residues (1.4%). It was postulated that PvPRP1 may be down-regulated during infection in order to be replaced by other tyrosine-rich PRPs or PRPs which function more effectively for defense, such as the PRPs P35 in bean and P33 in soybean (81). In contrast to the elicitor treatment response, in wounded bean hypocotyls, the PvPRP1 mRNA level was unaltered for 1 h, decreased to a minimal level at 2.5 h and then steadily accumulated to levels higher than that observed prior to wounding (443).

TRANSCRIPTIONAL AND POST-TRANSCRIPTIONAL CONTROL OF CELL WALL PROTEIN GENE EXPRESSION

In bean, extensin mRNA accumulation in response to both fungal elicitor and infection is at least in part attributable to increased rates of transcription, as measured by in vitro nuclear run-off experiments (286). After a 2 h lag, extensin transcription was stimulated with maximal transcription rates (about threefold stimulation above basal level) occurring 7 h after elicitor treatment. Elicitor caused a slower, less extensive but more prolonged stimulation of extensin transcription, in contrast to the rapid and transient transcriptional activation of two phytoalexin biosynthetic genes, PAL and CHS. Activation of extensin gene transcription was consistent with the accumulation of steady state extensin mRNA levels (between 4 and 10 h after elicitor treatment) (286).

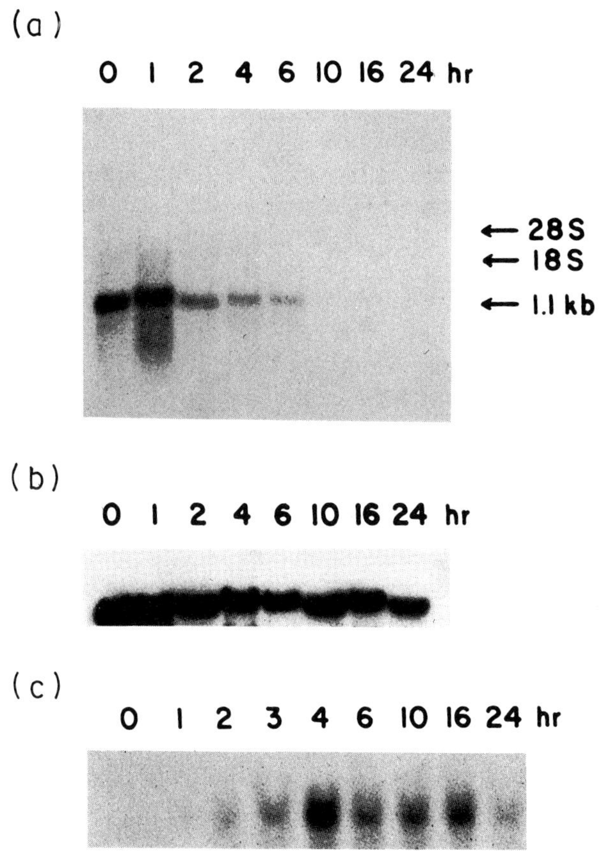

Fig. 3. Down-regulation of the PvPRP1 mRNA in elicitor-treated bean cells. Total cellular RNA was isolated at the times (in h) indicated after elicitor treatment. (a) The blot was hybridized with the ^{32}P-labeled Hinc II-Eco RI fragment of PvPRP1; (b) The same blot was washed and rehybridized with the ^{32}P-labeled constitutive clone H1. (c) Accumulation of mRNA encoding the phytoalexin biosynthetic enzyme chalcone isomerase (318) in the same set of RNAs.

The mechanism for down-regulation of the Hyp2.11 extensin mRNA by fungal elicitor is unknown. For PvPRP1 mRNA, while the steady state level of transcript declines to 5% of the control level 4 h after elicitor treatment, the transcriptional rate of these cells remains unchanged during this period, as measured by nuclear run-off assays (530). Thus, the decline in PvPRP1 mRNA abundance is caused exclusively by increased mRNA degradation. Two mechanisms for this mRNA degradation were considered by Mehdy and colleagues: 1) elicitor treatment triggers changes in preexisting factor(s) so that an increased rate of breakdown occurs or 2) *de novo* synthesis of factor(s) responsible for degradation occurs after elicitor treatment. Studies using actinomycin D, a transcriptional inhibitor, showed that PvPRP1 mRNA is very stable with an estimated half-life of about 60 h in unelicited control cells, compared with a half life of less than 20 min in elicitor treated cells. When cells were pretreated with actinomycin D to block transcription and then treated with elicitor, steady state PvPRP1 mRNA levels remained largely unchanged. Similarly, preincubation of cells with emitine or anisomysin, two protein synthesis inhibitors, prior to elicitor treatment also blocked RNA degradation, resulting in unaltered steady state PvPRP1 mRNA levels. These data indicate that de novo synthesized protein factor(s) are involved in the degradation of the PvPRP1 mRNA in elicitor treated bean cells (530).

Mehdy and colleagues hypothesized that two types of proteins are involved in the regulation of PvPRP1 mRNA abundance. The first type would bind to the mRNA in unstressed cells and stabilize the transcripts for normal developmental needs. The second type is synthesized de novo to bind and/or specifically degrade the transcript to achieve the down regulation, thus allowing other defense-induced cell wall structural proteins to be produced and integrated to reinforce the cell wall. The protein factor(s) could be either a target-specific RNase(s) or specific RNA binding protein(s) that facilitate the breakdown of the PvPRP1 mRNA. It is noteworthy that each of the 3' untranslated regions of the PvPRP1 mRNA as well as the Hyp2.11 mRNA contains two AUUUA motifs, which is a consensus motif found in numerous short-lived mRNAs in animal cells (82). It was recently reported that a homologue of the β subunit of farnesyltransferase in garden pea also contains two copies of the AUUUA motif in its 3'untranslated region and that the transcript levels decline in pea leaves when seedlings are grown in continuous white light (527).

PvPRP1 mRNA levels are probably also regulated by mechanisms which do not involve decreasing RNA stability. The phytochrome-mediated accumulation of PvPRP1 mRNA in light-grown bean hypocotyls provides one obvious example where other mechanisms, such as increasing mRNA stability and/or increased transcription, are probably at work (444).

IMMUNOCYTOCHEMICAL LOCALIZATION OF EXTENSINS IN INFECTED PLANTS

Generally, cell wall structural proteins exhibit a wide distribution with respect to specific tissues and cell types in healthy plants (451). A melon extensin antibody was used to localize extensin in melon and bean plants infected with bacterial or fungal pathogens. Extensin was found to accumulate markedly in wall-like papillae produced by plant cells in response to infection by fungi and bacteria. In the hypersensitive response, extensins markedly accumulate at walls and paramural papillae of living cells near the dead cells; this accumulation may help in restricting secondary infections to the necrotic tissue. However, the accumulation of diffusible phytoalexins may play an even more important role in restricting further infection (354). Similarly, marked accumulation of extensins was detected in tomato roots inoculated with compatible *Fusarium oxysporum* f. sp. *radicis-lycopersici.* Accumulation of extensins in primary walls of cortical and parenchyma cells and in secondary thickening of invaded xylem vessels seemed to be initiated after contact between the fungus and the plant cell wall (44).

ULTRA-RAPID CROSS-LINKING OF PRPs AND EXTENSIN IN RESPONSE TO STRESS

In cell wall preparations from cultured bean cells an abundant 35 kD PRP (P35), homologous to a 33 kD PRP in soybean, and a 100 kD putative extensin, are dominant SDS-extracted proteins revealed by polyacrylamide gel electrophoresis (PAGE). Soon after elicitor treatment, these two proteins are no longer observed following PAGE of the SDS-extracted fraction. The disappearance of these two proteins is thought to result from insolubilization via cross-linking rather than by degradation, since extensin and PRP antibodies still react strongly with cell walls of the elicitor-treated cultures (81). Bradley et al. (81) hypothesized that elicitor stimulates the disappearance of these proteins by an H_2O_2-mediated oxidative cross-linking, so that they are insolubilized in the cell wall and can not be extracted by SDS. Furthermore, H_2O_2 can stimulate the disappearance of these proteins in isolated cell walls, whereas the elicitor can not and requires intact cells to have the same effect.

Presumably, the elicitor causes the H_2O_2 burst and subsequent cross-linking by binding to a cell membrane receptor. The remarkably rapid disappearance of PRP and extensin is completed within 5-10 min and precedes all transcriptionally-activated defense responses. Such oxidative cross-linking also occurs in response to wounding and appears to be a normal developmental process (81).

SIGNAL TRANSDUCTION PATHWAYS LEADING TO GENE REGULATION BY FUNGAL ELICITOR

Bean suspension cultures challenged with fungal elicitor represent the best studied system for cell wall proteins involved in fungal infection. While the signal transduction pathway leading to the gene regulation in response to elicitor is still incomplete, a model can be made according to our current understanding (Fig. 4). First, it is likely that fungal elicitor binds to a receptor in the plasma membrane (108). It is also possible that fungal elicitor triggers the cell wall to release its own endogenous elicitor which binds to a plasma membrane receptor. In either case, we envision the elicitor receptor either directly or indirectly activating NAD(P)H-oxidase responsible for the observed burst of H_2O_2 production. H_2O_2 production was recently documented by using peroxidase-mediated fluorescence quenching of pyranine in elicitor- treated bean suspension cultures (Sharma and Mehdy, personal communication). This H_2O_2 burst triggers the ultra-rapid oxidative cross-linking of preexisting cell wall proteins such as P35 in bean and P33 in soybean and the 100 kD extensins in bean and soybean (81). H_2O_2 may also induce cross-linking between different classes of wall molecules including different classes of wall proteins and polysaccharides, presumably to construct a more impenetrable barrier.

Additionally, H_2O_2 may serve as a secondary messenger, responsible for the regulation of the levels of various defense-response mRNAs. Both elicitor and H_2O_2 treatment result in the accumulation of PAL and CHI mRNAs and in the decrease of the bean PvPRP1 mRNA; moreover, the effects of both treatments can be blocked with agents that disrupt H_2O_2 production (Sharma and Mehdy, personal communication). We speculate that H_2O_2 may indirectly act to stimulate transcription of up-regulated defense-response genes such as PAL and CHI [and also, we predict, PRPs and extensins] as well as the gene(s) encoding RNA binding protein(s) and/or target-specific RNase(s) responsible for the down-regulation of the PvPRP1 mRNA. We further speculate that RNA binding protein(s) will be discovered that serve to stabilize extensin

(and possibly other) mRNAs that exhibit sustained accumulation kinetics after elicitor treatment.
In addition to cross-linking, other post-translational modifications of cell wall proteins may be altered by fungal infection, but this is largely unexplored territory. Related to this idea, the interaction of such post-translationally altered wall proteins with other wall components may consequently be affected.

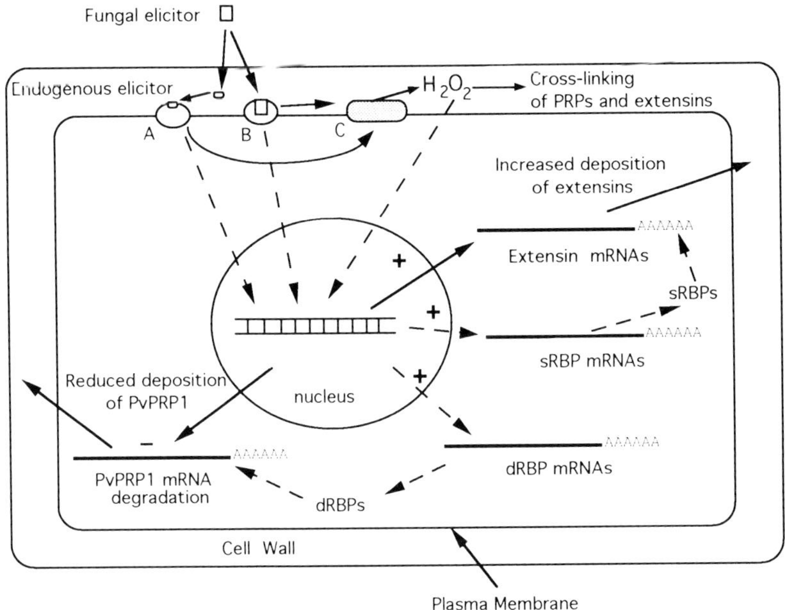

Fig. 4. Model of the regulation of PRP and extensin gene expression in elicitor-treated bean suspension cultures. Dashed lines indicate hypothetical pathways; + indicates induced transcription; - indicates down-regulation of RNA levels; A: receptor for endogenous elicitor; B: receptor for fungal elicitor; C: oxidase complex; dRBP: destabilizing RNA binding protein; sRBP: stabilizing RNA binding protein.

In summary, fungal infection brings about dramatic alterations in the plant cell wall, particularly with respect to the structural proteins which comprise it. These alterations occur at several levels: transcriptional activation of some wall protein genes, post-transcription control affecting RNA stability and turnover, and on the protein level via oxidative cross-linking. Increased deposition of certain types of extensins and PRPs for cross-linking and concomitant de-

pletion of "weak" wall proteins may lead to a new architecture that is more impenetrable for the fungal pathogens. Changes in extensin and PRP composition in the cell wall will also alter the net charge of the wall and thus affect ionic interactions among wall components and with the invading fungus. In addition, extensins are reported to serve as agglutinins to invading pathogens (320); this possibility has not been examined with respect to the PRPs yet. Clearly, changes in the cell wall structural proteins and hence remodeling of the cell wall occur and likely represent one aspect of an effective overall strategy in plant defense.

ALTERATION OF HOST CELL SURFACES BY MYCORRHIZAL FUNGI

Paola BONFANTE

Dipartimento di Biologia Vegetale, Università di Torino, I–10125 Torino, Italy

Mycorrhizae are widespread symbiotic associations established between the roots of over 90% of land plants and many soil fungi. Extensive investigations have demonstrated that many functions of both plant and fungal partners depend on the establishment of symbiosis. Mycorrhizae may play important roles in the success of a sustainable agriculture as well as in the structure of plant communities, in maintaining biodiversity, in improving plant nutrition and in allowing completion of the fungal life cycle (10, 239, 399). On the other hand, cellular and molecular analyses have revealed that the establishment of the mycorrhizal status depends on a number of events, including chemiotactic attraction, fungal adhesion, appressoria formation, host tissue penetration and colonization (74). Many of these events require that the two partners come into close contact and identify themselves as compatible or incompatible. This leads to a variety of physical cell-to-cell interactions that range from simple intercellular contacts to more complex intracellular ones. The cell surface of both partners is always directly involved.

Recent investigations have shown that the molecular components of the plant cell surface organize themselves into a complex structure consisting of cell wall, plasma-membrane and cortical cytoskeleton (415). According to this model, the cell wall is regarded as the major component of a dynamic extracellular matrix (258), not as an inert structural box within which resides the protoplast. The cell wall contains surface markers of plant development and components involved in communication. It can originate signal molecules which induce the production of soluble components involved in plant defense. The cell wall also can be modified in response to invading microorganisms by the deposition and insolubilisation of proteins or lignins (93 and quoted references).

This report will focus on the alterations of the host plant cell wall surface during mycorrhizal symbiosis. Changes occurring in the fungus during the symbiosis were recently reviewed (73, 76).

SIGNALLING AND NUTRITIONAL ROLES OF THE FUNGUS-PLANT INTERFACE

Plants and mycorrhizal fungi are always in physical contact, irrespective of the mycorrhizal types (arbuscular, ericoid, orchid, ecto- and ectendo-mycorrhizae; for definitions see, e.g., 212). The direct contact between plant and fungus through their cell surface leads to the establishment of an apoplastic *interface* structure. The role of this structure is to allow exchanges of signalling and nutritional molecules (76, 462). The term *interface* was first used by Bracker and Littlefield (80) to describe the contact area between plants and fungal pathogens, but it is now used in a more general sense to indicate morphological contiguity between walled organisms.

Two different types of interfaces are recognized in mycorrhizae, depending on whether the fungus develops among the host cells or penetrates them (Fig. 1).

In the first case, usually found in ectomycorrhizae, the walls of the two partners are in contact where the fungus grows over the epidermal cells, and where the hyphae develop between the cortical root cells producing the Hartig net. The same type of interface is also found in arbuscular mycorrhizae in the region where the intercellular hyphae spread along the root cortex. In the second case, which is found in all endomycorrhizae, the interface between the intracellular hyphae and the invaginated host membrane consists of extracellular matrix material of variable texture and composition.

Two major aspects of the interface area that influence its physiological role in mycorrhizae are the degree of alteration of the host wall and the molecular nature of the interfacial material. Whereas ultrastructural observations can only provide limited information to investigate these aspects, affinity techniques offer a bridge between morphological, molecular and biochemical observations. These techniques are based on the specific non-covalent binding between complementary sites on different macromolecules (such as antigen-antibody, sugar-lectin, substrate-enzyme and mRNA-cDNA) (204). Depending on the type of marker that visualizes the binding reaction, they can be used to locate molecular components in the different cell compartments or within tissues and organs. Affinity techniques have been shown to be powerful tools to understand many biological processes. In addition, the development of cryotechniques, such as high pressure and freeze-substitution, has strongly improved the preservation of cytology in plants and fungi involved in pathogen interactions, when samples are processed for electron microscopy (323). Many of these techniques have been used to investigate the plant-fungus interface in the different types of mycorrhizae.

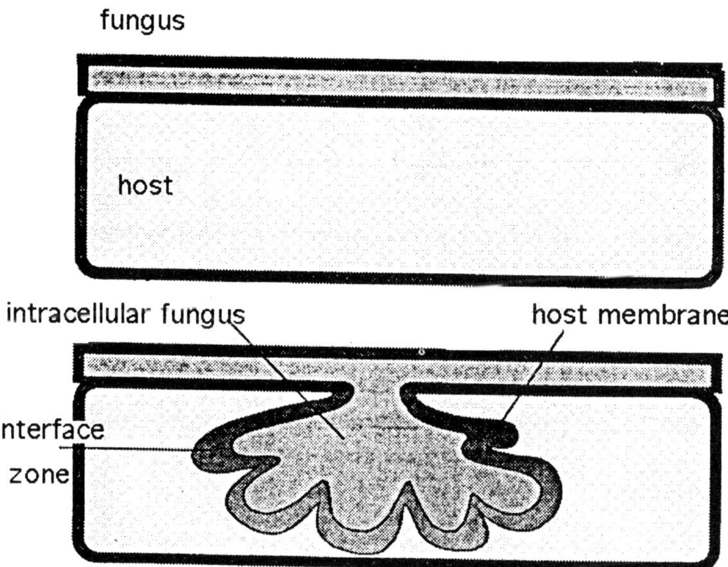

Fig. 1. The two different types of contact between the host plant and a mycorrhizal fungus are shown. Upper figure: intercellular contact; Lower figure: intracellular contact.

HOST WALL ALTERATION AT THE PLANT-INTERCELLULAR FUNGUS INTERFACE

Ectomycorrhizae

During the establishment of ectomycorrhizae the epidermal, cortical and root cap cells of Gymnosperms and Angiosperms come into contact with the hyphae of ectomycorrhizal fungi. The variety of host cell types involved in the formation of interfaces in the different regions of the root makes it difficult to determine to what extent the host wall is altered and to draw general conclusions. However, many developmental studies have considered ectomycorrhiza formation as a two-stage process (135, 145, 278). In the first stage, the hyphae form a contact with the root surface, whereas in the second stage a fungal sheath (the so-called Hartig net) is established, which encloses the epidermal and the root cap cells. During this second stage the Hartig net develops among the cortical cells. No evident alterations of the host cell wall have been described at the interface formed during the early stage of contact, with the exception of an abundant polysaccharidic mucilage on the surface of *Pinus strobus* roots (387). In contrast, evident changes have been reported during the later stage, when an *involving* layer is deposited between the host and the fungal wall. In this region, both symbionts lose their wall integrity. Information on the molecular nature of the involving layer is very limited. Labelling experiments on ectomycorrizae of *Corylus avellana* and *Tuber melanosporum* using wheat germ agglutinin, a lectin specific for N-acetylglucosamine residues, only labelled the fungal wall (76), indicating that chitin molecules of fungal origin do not take part in the formation of this layer. Nylund (353) also concluded, on the basis of its staining properties, that the involving layer is of host origin and is rich in pectic material.

The alterations observed in the cell wall of ectomycorrhizal plants are thought to facilitate the nutritional transfer between the partners. In addition, an active process involving membrane ATP-dependent H^+ pumps (462) might generate the driving energy for the uptake of phosphate by the host (76). Cytochemical observations have shown that ATPase activity occurs at the plasma membrane of pine root cortical cells during the first step of interaction with *Laccaria laccata*. The same activity is not present in the old cortical cells (289).

In conclusion, the host cell surface in ectomycorrhizal roots appears to be modified at the site of contact with ectomycorrhizal fungi. However, it will be necessary to investigate further the interface area with affinity techniques in order to establish the degree of host wall alteration and its molecular nature.

Arbuscular mycorrhizae

Arbuscular mycorrhizae (AM) are established with host plants representing about 80% of all plant species and ranging from ferns to monocots (212). Arbuscular fungi, belonging to the order of Glomales (509), undergo complex morphogenesis during mycorrhizal formation: they produce spores that germinate and develop a vegetative mycelium which, upon contact with the host root surface, produces appressoria. Hyphae originating from the appressoria initiate the intraradical phase, growing as intercellular hyphae or coils, and forming vesicles and highly branched structures called arbuscules.

During the infection process, some contacts between the host root cells and the AM fungus are intercellular (Fig. 1a) and comparable with those found in ectomycorrhizal associations (i.e., during the first contact phase and when the intercellular hyphae spread between the cortical cells). This raises the question whether the host cell surface is modified following the contact with the AM fungus. The morphology of appressoria has been recently investigated, and the behaviour of *Glomus mosseae* in the presence of host or non-host plants has been compared (186). Information on the host wall structure, however, is limited. Morphological observations have shown a localized thickening underlying the walls of epidermal cells (180). This feature is comparable – although reduced in extent – with the papillae that develop during plant-pathogen interactions. Larger wall thickenings have been found associated with the root epidermal cell walls of some pea mutants unable to form mycorrhizae, where the arbuscular fungi produce appressoria, but fail further penetration. Callose has been found in these mutants at the specific site of the thickenings (189).

On a host plant, the appressoria produce hyphae which are responsible for the infection process. Based on the different localization of cellulose, pectic, and phenolic molecules in the epidermal cell walls of leek, maize and pea, it has been hypothesized that the preferential site for fungal penetration is controlled by the physicochemical features of the host walls (74). For example, suberin and phenols constitutively present in exodermal cells act as a barrier to fungal penetration in many plants (74, 86). It appears that the colonization process of an arbuscular fungus in different host plants is influenced by the composition of the cell wall, and that fungal infection can be channelled through specific cell types. Striking changes, however, have never been detected in the composition of the host cell wall.

Once the fungus overcomes the barrier represented by the outer root layers, it develops hyphae in the intercellular spaces of the cortical tissue. Unesterified and esterified pectins, hydroxyproline-rich glycoproteins and cellulose have been localized in the wall of cortical cells in both monocots and dicots (see 74 for references) using affinity techniques with poly- and monoclonal antibodies, as well as purified enzymes (Figs. 2, 3). This is consistent with the structural models of cell walls recently proposed (93). However, the morphological analysis of cortical cell walls of leek roots have revealed a thin and simple structure in uninfected roots, which becomes swollen and polylamellate in the presence of the fungal infection (Fig. 4). Limited signs of degradation are often found on the middle lamella, although the fungus seems to spread intercellularly in preformed air channels (86). This suggests the presence of a weak but well defined alteration of the host cell wall caused by the intercellular fungus.

In conclusion, these studies point to the importance of the root cell surface during the first contact between plant and fungus in the formation of arbuscular mycorrhizae. In the meantime, they demonstrate that the spread of infection limited to the intercellular hyphae only causes limited alterations in the host cell wall.

HOST WALL ALTERATION AT THE INTERFACE BETWEEN HOST CELLS AND INTRACELLULAR FUNGI

Arbuscular, orchid and ericoid mycorrhizae are considered endomycorrhizae, because the fungus forms intracellular structures inside the root cells. In all these symbioses, as in some biotrophic pathogenic interactions, the host plasma membrane invaginates at the point of fungal penetration (Fig. 1) and proliferates around the growing fungus. A new apoplastic interface compartment with interfacial material is created (76). Although the components of the interface are comparable among the different endomycorrhizae – being

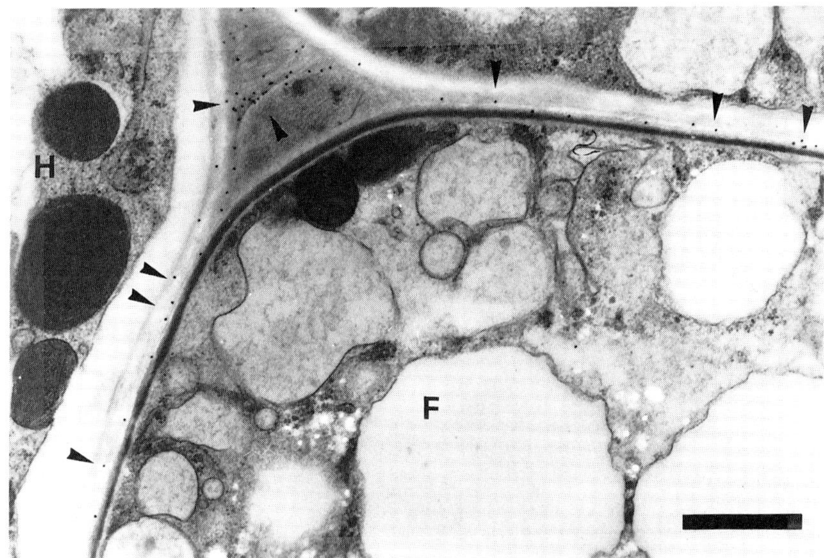

Fig. 2. Contact between an intercellular hypha of *Gigaspora margarita* (F) and the host (H) in high pressure-freeze substituted roots of clover. The middle lamella and the material at the plant cell junction are immunogold labelled using an antibody for pectins (arrows) (in collaboration with K. Mendgen). Bar corresponds to 1 µm.

represented by the host membrane, the interfacial material, the fungal wall and membrane – a detailed analysis is available only for arbuscular mycorrhizae.

The use of high pressure and freeze-substitution techniques supports the existence of the interfacial compartment. Samples processed by high pressure and freeze-substitution, refined for plant-pathogen interactions (323), demonstrate that the interfacial space exists, although of a reduced size compared to conventional fixation (about 80-100 nm) and that it is lined by a smooth membrane (Fig. 5). The material present in the space is continuous with the host wall but different in texture (376). The interfacial compartment is electron-dense and thick at the point of penetration, then it tapers and loosens around the active arbuscular branches

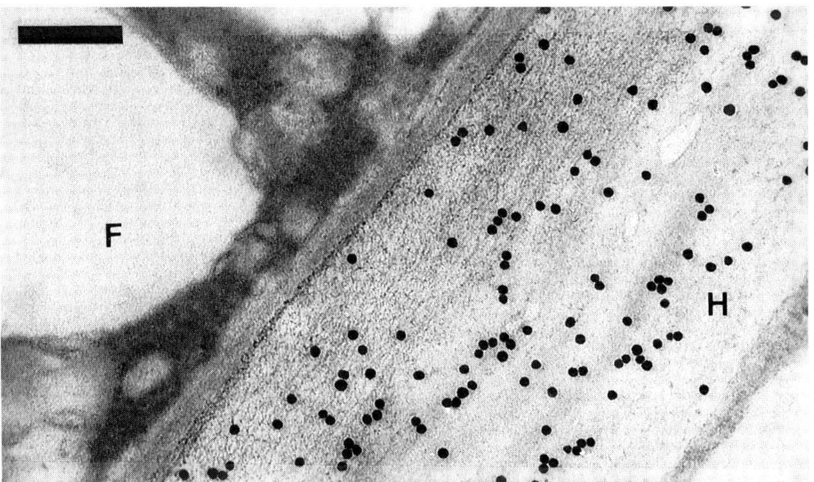

Fig. 3. Intercellular contact between the wall of a cortical cell in leek and hyphae (F) of *Glomus versiforme*. The host wall (H) is labelled by gold bound to CBH I, specific for (1-4)β glucans. Bar corresponds to 0.2 μm.

(Figs. 4, 5), to become thicker again around the collapsed arbuscular branches. Affinity techniques used on many plants (pea, maize, leek, ginkgo, clover) show that the interfacial compartment is a zone of high molecular complexity where glucans, pectins, hydroxyproline-rich glycoproteins (HRGPs) can be localized and identified as being of host origin (Fig. 5) (74). Many Golgi bodies are found closely associated to the interface compartment (Fig. 4, inset). In ericoid mycorrhizae, in contrast, the same probes have not clarified the nature of the interface material. The monoclonal antibodies JIM 5 and JIM 7, which are probes for polygalacturonans, failed to label the root epidermal cells of the host plant *Calluna vulgaris*. In addition, the enzyme cellobiohydrolase I (CBH I), produced by *Trichoderma* sp. and specific for (1-4) β glucans, labelled not only the host cell walls but also the wall of the fungus, *Hymenoscyphus ericae*. Due to this double labelling, it was impossible to define the origin of the material occurring in the interfacial compartment of ericoid mycorrhizae, which could be of either host or fungal origin.

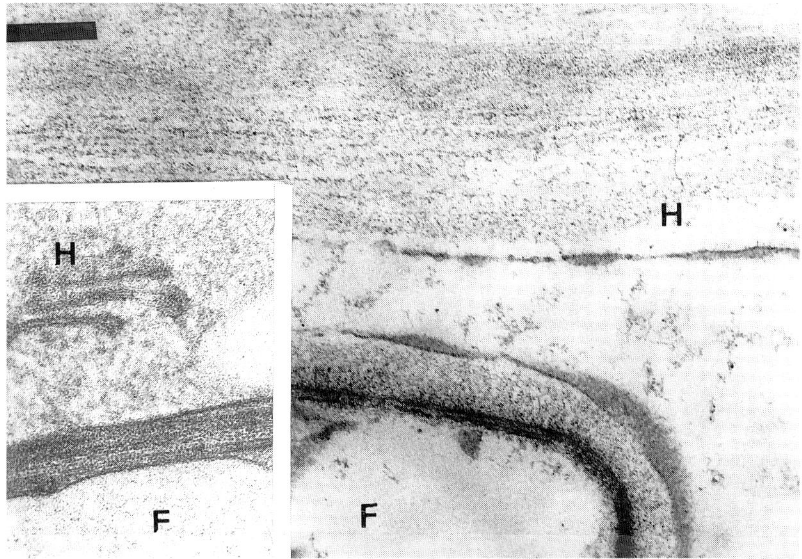

Fig. 4. Intracellular contact between an arbuscular branch (F) of *Glomus versiforme* and a leek cortical cell. The host wall is swollen and polylamellate while the interfacial material has a granular appearance. Bar corresponds to 0.25 μm. In the inset a Golgi body is shown.

The host membrane surrounding the intracellular fungus, often called periarbuscular membrane, is an important component of the plant-fungus interface. It is continuous with the peripheral plasma membrane, although important enzymatic differences exist. Gianinazzi-Pearson et al. (184) demonstrated that membrane-bound ATPases occur on the periarbuscular, but not on the peripheral plasma membrane. These cytochemical results are consistent with the hypothesis of a double nutrient flow occurring in mycorrhizae and mediated by an H+-ATPase pump. These results also underline the differences with the single way nutrient flow occurring in plant-pathogen interactions, where no ATPase activity was revealed on the perihaustorial membrane (181). The different ATPase distribution is mirrored by differences in membrane potential: experiments on mycorrhizal leeks have shown a stable hyperpolarization of the membrane (167). In contrast, in pathogenic associations depolarization of the host membrane is commonly observed (372): this may be indicative of an increased membrane permeability and of an ion efflux toward the fungus.

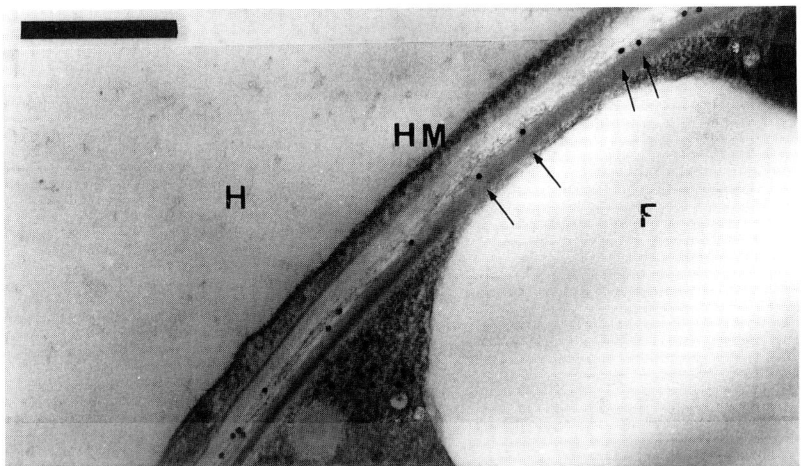

Fig. 5. Intracellular contact in high pressure-freeze substituted roots of clover colonized by *G. margarita*. An arbuscule (F) is surrounded by the invaginated host membrane (HM). The interface space is loosely labelled by gold granules after treatment with an antibody for pectin localization. Bar corresponds to 0.5 μm.

The relatedness between the peripheral plasma membrane and the invaginated periarbuscular one is shown by experiments where monoclonal antibodies, raised against membrane antigens from pea nodules, were used on mycorrhizal pea roots. As in the nodule, many of the probes targeted both the plasma membrane of root cells and the periarbuscular membrane of arbuscule-containing cells. However, other antibodies suggested some differences between the two types of symbiosis, as specific antigens (for example those recognized by the monoclonal antibody MAC 266) were more abundantly expressed during the mycorrhizal infection than during the nodule symbiosis (375).

The cyto-molecular dissection of the interfacial compartment in endomycorrhizae suggests that:

- the interface is an apoplastic compartment where many molecules are in common with the peripheral wall, although the morphology of the two compartments is different;
- the presence of the fungus can affect the assembly but not the expression of molecules which are usually present at the cell surface (membrane and cell wall molecules).

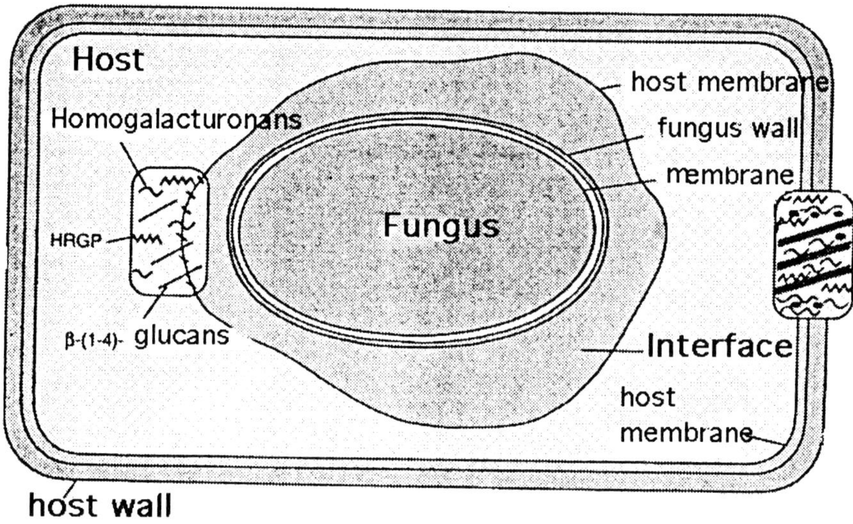

Fig. 6. Similarities and differences between the host wall and the interface as revealed by *in situ* labelling experiments.

Therefore, it is of particular relevance to understand why the structure of the interface is different from the peripheral wall. We propose that a root cell containing an endomycorrhizal fungus is involved in the *de novo* synthesis of cell wall-like material. Thus AM fungi would induce activation of specific genes in the host, which modifies its pattern of expression. Preliminary results have shown a pattern of HRGP mRNA expression in infected maize roots which is different from that described in the uninfected root. In addition to the usual labelling in the apical regions in the presence of the fungus, mRNA labelling has been localized in the cortical differentiated regions of mycorrhizal roots (Puigdomenech and

Bonfante, unpublished). The presence of cell wall molecules in the interface between the fungal and plant cells recalls therefore the situation of meristematic cells which are involved in the construction of their new peripheral cell wall. In the differentiated cells, however, and differently from what happens in a meristem, the presence of the endomycorrhizal fungus causes a physical and/or enzymatic interference with the usual mechanism of wall assembly. This probably leads to the morphological differences existing between the peripheral wall and the interface zone (Fig. 6).

THE MOLECULAR BASES OF ALTERATIONS OF HOST CELL SURFACES

The alterations shown by the host cell surface during the process of mycorrhiza formation include slight wall changes and the creation of a new compartment (the interface) where new membrane and wall molecules are laid down. These modifications are the results of the non-aggressive strategies which characterise mutualistic symbioses. We do not know how the balance between host and fungus is reached and how mycorrhizal fungi avoid elicitation of the host plant defence response. Some hypotheses involving specific sensor genes or nutritional and hormonal effects have been suggested to explain the limited expression of plant defence genes during the symbioses (189, 284). We believe that a low or finely regulated production of fungal cell wall hydrolytic enzymes is also crucial to avoid a high level of elicitation. Cell wall degrading enzymes are produced to a different extent by pathogens depending on their nutritional strategies. Their role as determinants of pathogenicity and elicitors has been well established (119). In contrast, information on the production of hydrolytic enzymes by mycorrhizal fungi is very limited, and it has to be verified whether the penetration events (for example during arbuscule formation) result from mechanical and/or enzymatic processes. As biotrophic fungi, they must colonize the host tissues still maintaining the host cell viability, as their survival depends on the host being alive. There is morphological evidence that the alterations of the host wall are limited, although loosening of the middle lamella can be observed.

Some ectomycorrhizal fungi produce extracellular polygalacturonases (252), whose activity in culture is delayed and very low in comparison with the activity of the necrotrophic *Botrytis fabae*. Similar results were obtained by comparing different ericoid strains with a pathogenic *Fusarium* sp. (97). Detailed investigations on an ericoid strain, PS IV, indicate the production of different forms of polygalacturonases (PG): an exoPG and an endoPG have been identified, both of them characterized by acidic pI (4.2. and 5.3, respectively) (373). In constrast, very low PG activity has been measured in extracts of spores of AM fungi (178) or in the soluble extract of mycorrhizal leek roots. In the latter case, PG activity is comparable to extracts from uninfected roots (Peretto et al., unpublished). *In situ* immunolabelling with an antibody raised against a PG from *Fusarium moniliforme* allowed us to detect some labelling around the arbuscular branches, suggesting that AM fungi may also produce polygalacturonase, although in limited quantity.

CONCLUSIONS

Strategies of colonization by mycorrhizal fungi are usually nonaggressive. The first morphological evidence for this is that alterations of host cell surface are limited in all the mycorrhizal symbiosis so far analyzed. The most important change is caused by the intracellular mycorrhizal fungi, which produce the interface compartment. This, however, does not represent a drawback from a functional point of view: nutritional exchanges are guaranteed by the absence of cell wall molecules which could act as an apoplastic barrier. The interface compartment prevents a direct contact between the fungal wall and the host cytoplasm. This is probably one of the successful keys for mutualistic symbioses.

HISTOLOGICAL STUDIES OF HOST PENETRATION AND COLONIZATION BY ENDOPHYTIC FUNGI

Jeffrey K. STONE[1], Olivier VIRET[2*], Orlando PETRINI[2] and Ignacio H. CHAPELA[3]

[1]Department of Botany and Plant Pathology, Oregon State University, Corvallis, OR 97331-2902, USA
[2]Microbiology Institute, ETH–Zentrum, CH-8092 Zurich, Switzerland
[3]3144, O Street W, Washington, D.C. 20007, USA

Traditionally, the presence of fungi in the foliage of plants has been related to disease symptoms followed by the decay of the plant tissue. In recent years, however, fungal infections of healthy, symptomless leaves of a variety of host plants have been reported. These infections have been termed endophytic, emphasizing their symptomless nature. It is now generally accepted that such infections are widespread in living plants (377, 379). Objectives of research on endophytic fungi during the past ten years have been documentation and quantification of symptomless infections in healthy plants, studies on the species composition of endophyte assemblages, and the evaluation of the nature of the host–endophyte symbiosis.

Evidence for the presence of endophytic fungi in many plants has been provided solely on the basis of isolations of the fungi from surface-sterilized tissue. From ecological and floristic analyses of a broad range of host species, it is apparent that endophytic fungi comprise a unique and complex ecological group distinct from obligate parasites and saprobes, although the ecological role of many endophytes is not well understood.

Ecological studies have demonstrated differences in infection frequencies related to age of foliage (160, 474), site microclimate (378, 454), and host genotype (493). Because they have been little investigated, endophyte assemblages of higher plants have proved to be rich in novel taxa.

Despite the growing number of isolation studies on endophytic fungi in the last few years, a paradoxical situation has developed

* Present address: Station Fédérale de Recherches Agronomiques de Changins, CH–1260 Nyon, Switzerland

where only few anatomical studies have confirmed the presence of these fungi within the plant tissues. Such studies are desirable not only to document the biological and cellular interactions between hosts and endophytes, but also to provide more accurate estimates of endophyte infection levels (91, 381, 475). Several technical obstacles have hindered this area of research. Because of the diffuse and sporadic nature of many endophyte infections, routine sectioning methods are not suitable for examining the large amounts of host tissue necessary to locate infections. Clearing and differential staining of entire leaves or other host tissues has provided information on endophyte colonization for some hosts at the light microscope level. Clearing procedures, however, can remove or alter superficial fungal infection structures, making this approach unsuitable for some studies. When endophyte infections can be detected both microscopically and by isolations, matching an observed colonization pattern with a specific fungal taxon may not be possible in the absence of immunospecific staining or critical diagnostic characters. Electron microscopy has been used successfully to overcome some of these obstacles (226, 382, 383, 384, 385, 386, 455, 475, 482, 507, 508).

Using immunological methods, Suske and Acker (482) have identified inter- and intracellular hyphae observed in the mesophyll of symptomless needles of Norway spruce (*Picea abies*) as *Lophodermium piceae*. *Rhabdocline parkeri*, an endophyte of Douglas-fir (*Pseudotsuga menziesii*), occurs as discrete intracellular infections which are limited to a single epidermal cell (474, 475). Intracellular infections similar to those of *R. parkeri* have also been observed for *Stagonospora innumerosa* infecting *Juncus effusus*, *Drechslera* sp. infecting *J. bolanderi*, and an undetermined species infecting *J. bufonius* (91). *Phyllosticta abietina* forms limited intercellular infections in the mesophyll of Douglas fir and *Abies* spp. that originate from penetration between epidermal cells (473). *Phyllosticta concentrica*, however, colonizes needles of *Taxus brevifolia* as a network of subcuticular hyphae growing between the margins of epidermal cells, but not colonizing internal tissue (476). Hyphae of some leaf-inhabiting endophytes, such as *Discula umbrinella*, apparently form subcuticular infections that invade stressed or senescent host tissues (506). Other leaf-inhabiting fungi, such as *Alternaria alternata* and *Cladosporium cladosporioides*, reside in substomatal chambers, from which limited intercellular colonization of the mesophyll can be observed (91).

From the few model endophyte systems that have been examined histologically, some consistent patterns have emerged that help to differentiate general categories of host-endophyte associations.

Grass endophytes form highly adapted, systemic associations with their hosts (20, 91), are mainly seed-borne (115) and are either transmitted by way of the seed coats and thence into the germinating seedling or by entering the embryo sac and infecting the embryo itself (384, 385, 386). Taxonomically, the grass endophytes are members of the Balansiae in the Clavicipitaceae. Because of the economic importance of pasture grasses, there has been an increasing amount of research in recent years on their endophytes and the host/fungus relationship. Most studies, however, have focused on the ecology and physiology of the association and surprisingly little work has been carried out on the histological aspects of the symbiosis (226, 381, 382, 383, 384, 385, 386, 455). In the associations between grasses and clavicipitaceous endophytes, typified by *Epichloë typhina*, host tissue is extensively colonized by intercellular hyphae (e.g., 114, 226, 382, 386, 455, 516). The extensive intercellular colonization of the host plant by the toxigenic *Acremonium* anamorph of *E. typhina* accounts for the distribution and concentration of fungal alkaloids in infected plants, and their well documented effects on vertebrate and invertebrate herbivores. This and a variety of other benefits to the host (116, 404) are frequently cited as the basis for a coevolved mutualism betwen *E. typhina* and its hosts.

Associations between host grasses and *E. typhina*, however, encompass a range of interactions from symptomatic associations with complete sterilization of the host plant, "choke disease", to partial host sterilization with fertile seeds produced that are infected with *E. typhina*, to completely asymptomatic endophytic associations with complete suppression of fungal sporulation on the host (515). Indeed, the ability of the latter endophytes to reproduce on their hosts has apparently been lost completely and the host plant and endophyte essentially function as a single organism (456).

Ecologically and biologically, the clavicipitaceous (and few non-clavicipitaceous: cf. 517) endophytes of grasses appear to be fundamentally distinct from the endophyte associations of non-grass hosts that have been examined histologically. These are primarily colonisers of woody plants, perennials, evergreen conifers, and hosts with long-lived or evergreen foliage. Taxonomically, endophytes of these hosts represent a broad range of genera from several orders and families of ascomycetes or from anamorphic form-genera. In contrast to the extensive hyphal colonization of grasses by clavicipitaceous endophytes, patterns of endophyte colonization in other hosts are much more limited. These fungi cause very restricted, non-systemic infections in healthy tissues and are transmitted from host to host by spores. In general, the infections appear to remain physiologically quiescent and limited with respect to host colonization of healthy or uninjured tissue. Such endophytes resume active colonization at the onset of senescence or after host injury and eventually sporulate.

"Non-grass" endophyte associations have been compared to those of the grasses and mutualistic interactions are often presumed, but a benefit to the host in such associations is less readily apparent. Carroll (94) has advanced the hypothesis of acquired variable host defense. In this scenario, innate host defense metabolites are supplemented with endophytic fungal metabolites toxic to insect herbivores. The added variability of fungal metabolites from recombination and multiple infections of the same host organs by different endophyte chemotypes would present an obstacle to insect herbivores in overcoming innate host defenses. This type of supplemental defense would be of particular benefit where the host is long-lived and there is a large disparity between insect and host reproductive cycles. However, because of the small amount of living host tissue colonized by these endophytes, it is unlikely that significant concentrations of fungal metabolites would exist in healthy undamaged tissue.

Several species of endophytic fungi are known to be taxonomically closely related to virulent pathogens of the same or related hosts (94) and may reflect recent speciation. In pine needles, the endophytes *Lophodermium pinastri* and *L. conigenum* belong to the same genus as the pathogen *L. seditiosum* (328). *Rhabdocline weirii* and *R. pseudotsugae* (364) are taxonomically closely related to *Rhabdocline parkeri*, a symptomless endophyte of Douglas fir. The *Meria* anamorph of *R. parkeri* is morphologically virtually identical to, and probably a close relative of, *Meria laricis*, a defoliating pathogen of *Larix* spp. (447). These close relationships between endophytic and pathogenic species within a genus imply that divergence has occurred relatively recently, and suggest that similar genetic mechanisms which maintain equilibria between hosts and pathogens also operate in host-endophyte associations.

CASE HISTORIES OF HOST PENETRATION AND COLONIZATION

Rhabdocline parkeri on Douglas-fir

Rhabdocline parkeri is a very common and widespread endophyte in the foliage of Douglas-fir, its only known host, in western Oregon and Washington. Despite its broad distribution and common association with the dominant forest tree species, it remained undiscovered until the survey of coniferous tree endophytes published by Carroll and Carroll (95), and remained formally undescribed for another ten years (447). Infections of *R. parkeri* in Douglas-fir foliage are extremely cryptic and its sporulating states are inconspicuous. During the endophytic phase, infections are limited to a single epidermal cell and each infected cell corresponds to a separate infection locus. Many separate, individual infections can occur on a single needle (474, 475). Morphologically, infected needles appear entirely healthy and without any external symptoms of infection. Upon germination, conidia of *R. parkeri* become two celled, one cell of which becomes pigmented and forms a lateral appressorium (Fig. 1A, B). Invasion of the epidermal cells is accomplished by direct penetration of the host epidermal cell wall via the appressoria and a fine infection hypha. The inner wall of the appressorium is contiguous with the outer wall of the infection hypha, and terminates a short distance into the host cuticle. Below this point the infection hypha has a single, very thin wall (Fig. 1D). As a result of invasion, the host epidermal cell is killed, and the invading hypha eventually enlarges to occupy the entire cell lumen. However, no colonization of adjacent, contiguous cells occurs in healthy tissue. The mechanism for limiting the infections to single cells is not known. Transmission electron microscopy of infected cells shows the intracellular hyphae to contain large quantities of lipid, numerous mitochondria and peroxisomes, indicating metabolic activity (Fig. 1C, E). No ultrastructural or histochemical changes indicative of host defense responses have been observed in adjacent cells. Infected cells contain remnants of cellular organelles and osmiophilic substances identified as condensed tannins (475). Initial infection of needles occurs in their first growing season, and needles may be repeatedly reinfected during the course of their lifetime, resulting in logarithmic increase in the densities of infected epidermal cells with needle age. A relatively minor proportion of the entire epidermis is infected however, estimated at below 5% in even the most heavily infected tissue (474). The accumulated infections in older needles may represent a diversity of genetically distinct individuals. McCutcheon et al. (314) have reported as many as six unique genotypes of *R. parkeri* within a single needle. At the onset of host senescence, i.e. in attached but yellowing needles, active growth of *R. parkeri* resumes with the invasion of adjacent cells by haustoria. Colonization proceeds to contiguous cells until needle abscission, when conidiophores of the *Meria* anamorph emerge from the stomata of abscised needles. The ascomata are eventually produced on fallen needles in mid-winter, as a small flap of epidermis lifts to reveal ascomata of approximately 0.25 mm diameter (447). The normal lifespan of Douglas-fir needles can exceed seven years and thus the endophytic phase of *R. parkeri* may commonly exceed five years (474).

Discula umbrinella on Fagus sylvatica

Sieber and Hugentobler (453) have reported *Discula umbrinella*, the anamorph of *Apiognomonia errabunda*, as a symptomless endophyte from virtually all beech leaves sampled. The ecology of *D. umbrinella* has been studied extensively (335, 453, 494), but until recently the mechanisms underlying penetration and colonization of the leaf tissues by this fungus have remained unknown. As in the *R. parkeri*/Douglas fir symbiosis, several infections can occur on the same leaf and several *Discula* genotypes can be detected within a single leaf (206). The first germination-related structures are produced by *D. umbrinella* two hours after application of the spore suspension to the leaf surface. Non-germinating or germinating conidia with long superficial hyphae, conidia bearing distinct, appressorium-like swellings at the ends of the germ tubes, and halo formation around the tip of germ tubes without appressoria, can be observed on the beech leaf surface. Apparently only conidia forming appressoria-like structures or halos are able to penetrate the host. Conidia form the same infection structures on the abaxial as on the adaxial surface of the beech leaf, but infection on the abaxial side yields consistently higher colonization (506). On the abaxial side, however, direct penetration through the stomata has been rarely observed and germ tubes most often pierce the outer edges of the guard cells.

Penetration of the host by *D. umbrinella* can occur either by means of a fine penetration hypha that invades the epidermal cell directly, piercing both the cuticle and epidermal wall (Fig. 2A), or alternatively, via limited subcuticular fungal hyphae that lift the cuticle and grow between the cuticle and the epidermal cell wall at places which

---

Fig. 1. *Rhabdocline parkeri* in Douglas-fir needles. **A**. Conidium germination on the surface of needle (light micrograph). Conidium has become 2-celled and has formed a lateral appressorium (arrow). **B**. Conidium germinating on needle surface (SEM); hyaline cell has collapsed, pigmented cell bears a lateral appressorium (A). **C**. Intracellular hyphae in epidermal cells (light micrograph). **D**. Penetration of host cuticle by germinating conidium (TEM). Note termination of outer germ tube wall. **E**. Intracellular hyphae in epidermal cell. Host cell contains osmiophilic material and remnants of organelles. Fungal cell contains nuclei, mitochondria, peroxisomes, lipid bodies. Scale bars: **A**, **B**, **C**, 10 µm; **D**, **E**, 1 µm.

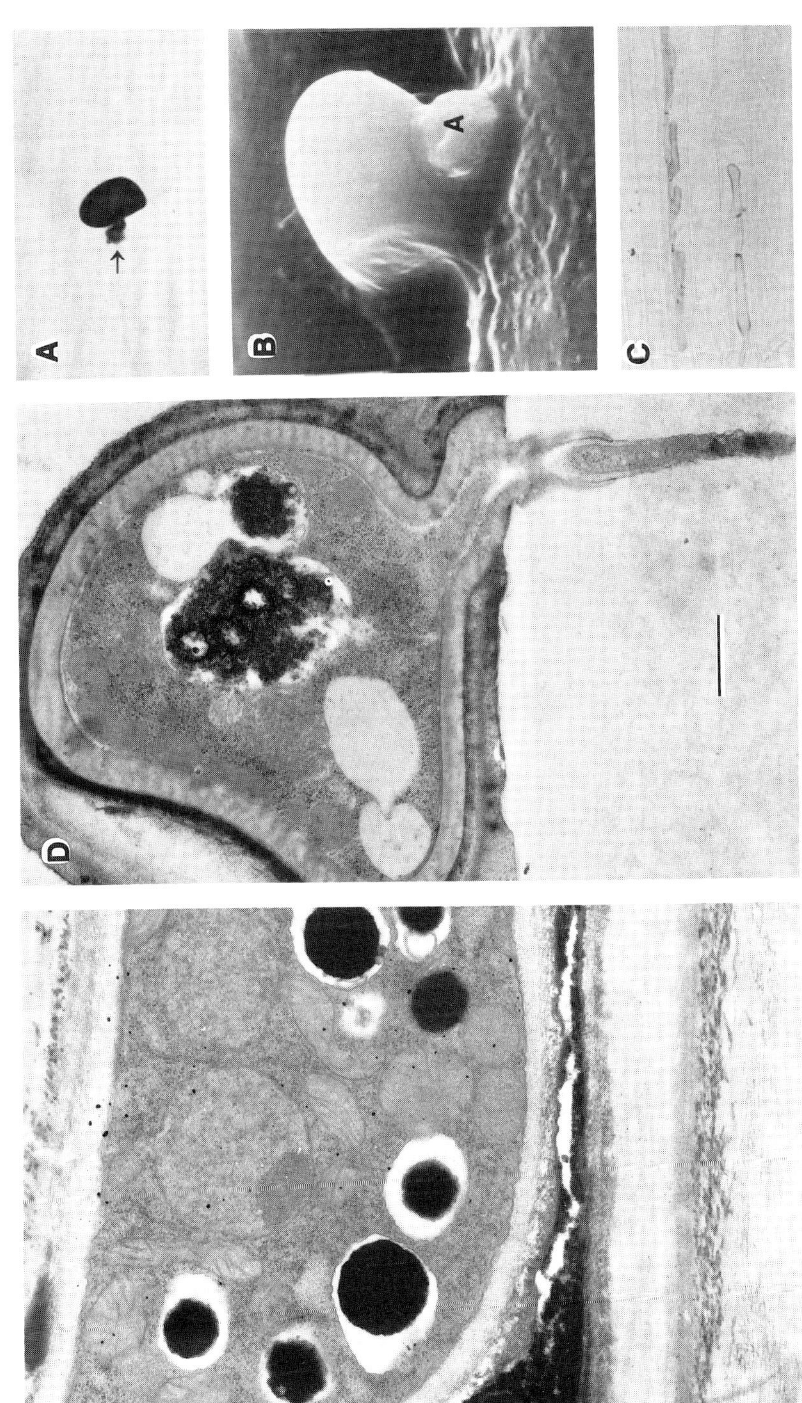
A
B
A
C
D
E

correspond to the halos observed on the host surface (Fig. 2B). The pattern of conidium germination, appressorium formation, and host penetration by *D. umbrinella* on beech is very similar to that described for the foliar pathogen *Gloeosporium platani*, a close relative of *D. umbrinella*, on sycamore (*Platanus*) leaves (438). In contrast to *G. platani*, in which almost all conidia form appressoria at the end of the germ tubes, only a small percentage of *D. umbrinella* conidia germinate with appressorium or halo formation. Evidence of enzymatic degradation of the cuticle and epidermal cell wall at the site of penetration can be seen in the TEM (Fig. 2A). The epidermal cell walls beneath the subcuticular hyphae appear at least partly digested, the cytoplasm has lost internal organization and an accumulation of osmiophilic, electron dense substances is apparent (Fig. 2B). This material, possibly condensed phenolic compounds, can be observed after protease-detergent treatment (231) of infected sections, but is not present in uninfected sections (507). Necrosis of the epidermal cell and accumulation of osmiophilic substances are probably related to a hypersensitive host response to infection. Necrotic lesions are usually restricted to sites of spore deposition. Structural organization of the mesophyll cells underlying lesions seems initially not to be affected, although the parenchyma cells show signs of incipient degeneration (507) and eventually also become necrotic. Similar reactions were observed in epidermal cells below appressoria of *Colletotrichum graminicola* on maize and oat leaves (391, 392). Limited colonization of the host tissues has been observed in infection experiments carried out with the oak leaf endophyte *D. quercina* by Wilson (520). In detached beech leaves, however, *D. umbrinella* grows progressively into the parenchyma through the epidermal cell wall and colonizes mainly the large intercellular spaces of the mesophyll, although at later infection stages intracellular growth is occasionally seen. Intracellular penetration and colonization are accompanied by decomposition of the cell wall (507).

Fig. 2. Beech leaf infected by *Discula umbrinella*. **A**. Penetration after appressorium formation. Note cuticle degradation at site of penetration (arrow) and epidermal cell wall degradation after subcuticular growth (arrowhead). **B**. Electron dense material in the epidermis cell underneath a subcuticular hypha. Ap, appressorium; c, cuticle; E, epidermis cell; eW, epidermal cell wall; h, hypha. Scale bars: 1 µm.

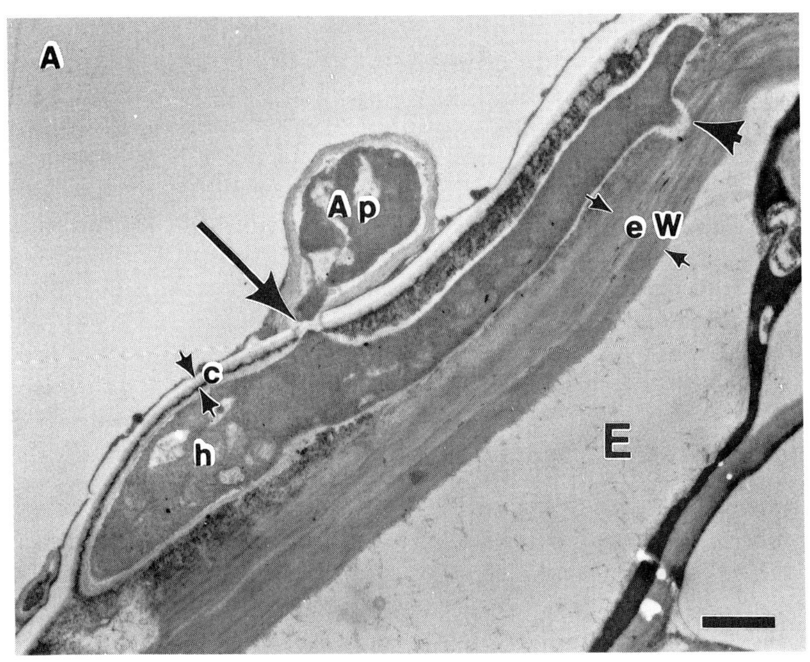
A
Ap
e W
c
h
E

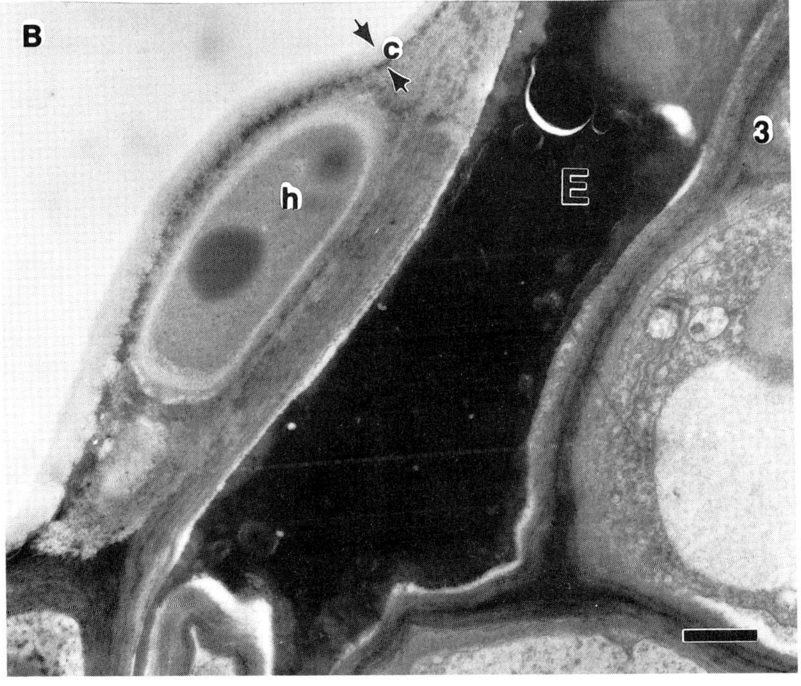
B
c
h
E
3

HOST SPECIALIZATION BY ENDOPHYTIC FUNGI

Successful establishment of infections of biotrophic, parasitic fungi on a host is a complex process, and host plants have evolved a variety of structural and chemical means to prevent or limit the occupation of tissue by parasitic fungi. These mechanisms serve to exclude the majority of opportunistic fungi, so that only highly adapted species are able to breach the defenses of a narrow range of hosts.

Implicit in the ability to overcome the complement of host defenses is a high degree of host adaptation. Appressoria and other specialized infection structures have evolved in specialized parasites to effect host penetration (8, 460). A combination of enzymatic and mechanical effects has often been observed (e.g., 233) in cases of direct penetration of host cell walls by fungi. Host responses to invasion include accumulation of refractory substances, papilla formation, cell necrosis (hypersensitive reactions) and synthesis of phytoalexins.

Although information on mechanisms of host penetration and colonization by endophytic fungi is available for very few models (91, 382, 474, 475, 482, 507, 508), the available evidence indicates that endophytic fungi have evolved host specific recognition and penetration mechanisms similar to those of other specialized fungal symbionts. Toti et al. (495) have demonstrated a differential attachment of *D. umbrinella* conidia to host surfaces and Chapela et al. (104) have shown that eclosion in *Hypoxylon fragiforme* is induced by the presence of a mixture of two monolignols, faguside and syringuside, present in the host tissues. Differentiation of appressoria precedes successful host invasion in both *Rhabdocline parkeri* and *Discula umbrinella*. Appressorium formation in *D. umbrinella* was observed only on the host surface and not on non-hosts or artificial surfaces (508). Viret et al. (508) have also described a physical contact with a surface is crucial for the germination of *D. umbrinella* conidia. Extracts of leaves of host plants enhanced germination but not differentiation of infection structures in either the conidial suspensions or on artificial and non-host surfaces. Pigmentation of *Meria laricis* and *R. parkeri* conidia, the first stage of appressorium differentiation and a prerequisite for successful infection, is also stimulated by host extracts (313).

Haustoria are structures common to both endophytic fungi and other specialized parasites and also indicate a high degree of host adaptation. They have been reported in *R. parkeri* (475) and *D. umbrinella* (508), and intracellular hyphal swellings have been described for *Lophodermium piceae* (482). Ultrastructural differences between haustoria of *D. umbrinella* and *R. parkeri* and those of other biotrophs have been noted, such as the absence of the typical neckband ordinarily present in other biotrophic parasites (303), which may reflect the relatively brief, transitional biotrophic phase of these two endophytes (475). In clavicipitaceous endophytes haustorium formation has only rarely been observed but Philipson (382) has reported that the *Gliocladium*-like endophyte of ryegrass (*Lolium perenne*) develops intracellular hyphae in senescent blades that penetrate the host cell walls with narrow perforation hyphae. A membrane that separates the penetrating structures of the endophyte from the host cytoplasm, described by Philipson and Christey (386) for the clavicipitaceous endophyte of *Lolium perenne*, has not been observed in either *D. umbrinella* or *R. parkeri*.

Patterns of endophyte colonization of host tissue differ markedly from those of most other fungal symbioses, whether mutualistic or antagonistic, in that despite the variety of mechanisms of host invasion, very little host tissue is colonized following initial invasion. Active host defenses triggered by initial invasion are probably responsible for restriction of endophyte colonization, however, little evidence for such responses exists. Cabral et al. (91) have reported the accumulation of phenolics, pigmentation of infected cells and the formation of papillae in cells adjacent to infection sites of *Stagonospora innumerosa* in *Juncus effusus*. These infections, like those of *R. parkeri* on Douglas-fir, are restricted to single epidermal cells. Accumulation of condensed phenolics in infected cells was reported for Douglas fir cells invaded by *R. parkeri* (475). A hypersensitive reaction of the plant against *D. umbrinella* is at least suggested by the accumulation of electron-dense compounds observed in the epidermal cells after penetration (507).

Endophytic fungi are as highly specialized on their hosts as pathogenic fungi, but in contrast they colonize internal host tissue without manifest symptoms. The endophyte life history strategy is characterized by early establishment in living host tissue. A quiescent biotrophic phase follows that may be of long duration, although the degree of host colonization may be very limited. The biotrophic phase is followed by a saprotrophic phase coinciding with senescence or injury to the host during which extensive colonization and sporulation occur. This strategy ensures early occupation and possession of the nutritional resources before colonization by opportunistic fungi can occur. Fungi alternately exhibiting both biotrophic and saprobic nutritional modes have been termed hemibiotrophs (120, 300). This distinct life history strategy implies a close host-parasite association, through which evolutionary selection can act to modify the nature of the symbiosis toward the extremes of mutualism or antagonism, as discussed by Carroll (94).

Although represented by relatively few detailed studies, histological information has provided a basis for understanding the unique life history strategy of endophytes. Histological studies of model host-endophyte systems documenting the localization and distribution of endophytic fungi in host tissue, and the mechanisms of host penetration and colonization have enabled comparisons between host-endophyte and other antagonistic and mutualistic plant-fungus symbioses. It is clear from these studies that endophytes represent a heterogeneous biological group. Common attributes of endophytic fungi are: (1) they are internal, at least subcuticular, in contact with and deriving nutrition from living host tissue, (2) they establish at least a transitory biotrophic nutritional relationship with their hosts, and (3) infected host tissue remains symptomless, i.e. disease free during its lifetime. Further study of endophytic fungi will surely provide new information on the nature of fungal infection processes, host responses to infection, mechanisms of limiting fungal colonization, and the interchange of metabolites in host-parasite relationships.

LITERATURE CITED

1 Adamovics, J. A., Johnson, G., and Stermitz, F. R. 1977. Ferulates from cork layers of *Solanum tuberosum* and *Pseudotsuga menziesii*. Phytochem. 16:1089–1090.

2 Adaskaveg, J. E. 1992. Defense mechanisms in leaves and fruit of trees to fungal infection. Pages 207–245 in: Defense Mechanisms of Woody Plants against Fungi. R. A. Blanchette and A. R. Biggs, eds. Springer–Verlag, Berlin.

3 Adaskaveg, J. E., Gilbertson, R. L., and Blanchette, R. A. 1990. Comparative study of delignification caused by *Ganoderma* species. Appl. Environ. Microbiol. 56:1932–1943.

4 Agosin, E., Blanchette, R. A., Silva, H., Lapierre, C., Cease, K. R., Ibach, R. E., Abad, A. R., and Muga, P. 1990. Characterization of palo podrido, a natural process of delignification in wood. Appl. Environ. Microbiol. 56:65–74.

5 Agrawal, V. P., and Kolattukudy, P. E. 1978. Purification and characterization of a wound–induced w–hydroxy fatty acid: NADP oxido–reductase isolated from potato tuber disks (*Solanum tuberosum* L.). Arch. Biochem. Biophys. 191 :452–465.

Mechanism of action of a wound-induced ω-hydroxy fatty acid: oxidoreductase from potato tubers (*Solanum tuberosum* L.). Ibid., 191:466–478.

6 Aist, J. R. 1976. Papillae and related wound plugs of plant cells. Annu. Rev. Phytopathol. 14:145–163.

7 Aist, J. R., and Bushnell, W. R. 1991. Invasion of plants by powdery mildew fungi, and cellular mechanisms of resistance. Pages 321–345 in: The Fungal Spore and Disease Initiation in Plants and Animals. G. T. Cole and H. C. Hoch, eds. Plenum Press, New York.

8 Aist, J. R., and Israel, H. W. 1977. Papilla formation: timing and significance during penetration of barley coleoptiles by *Erysiphe graminis hordei*. Phytopathol. 67:455–461.

9 Alder, E. 1977. Lignin chemistry–past, present and future. Wood Sci. Technol. 11:169–218.

10 Allen, F. 1992. Mycorrhizal functioning. Chapman and Hall, New York London.

11 Anderson-Prouty, A. J., and Albersheim, P. 1975. Host pathogen interactions. VIII. Isolation of a pathogen synthesized fraction rich in glucan that elicits a defense response in the pathogen's host. Plant Physiol. 56:286–291.

12 Atalla, R. H., and Agarwal, U. P. 1984. Raman microprobe evidence for lignin orientation in the cell walls of native wood tissue. Science 227:636–638.

13 Atkins, E. D. T. 1979. Pages 311–316 in Applied Fibre Science, Vol. 3, F. Happy (ed.) Academic Press.

14 Atkins, E. D. T 1992. Three-dimensional structure, interactions, and properties of xylans. Pages 39–50 in: Xylans and Xylanases. J. Visser et al., eds. Elsevier Science Publishers.

15 Ayers, A. R., Ayers, S. B., and Eriksson, K. E. 1978. Cellulose oxidase, purification and partial characterization of a hemoprotein from *Sporotrichum pulverulentum*. Eur. J. Biochem. 90:171–181.

16 Baayen, R. P. 1986. Regeneration of vascular tissues in relation to *Fusarium* wilt resistance of carnation. Neth. J. Plant Pathol. 92:273–285.

17 Baayen, R. P. 1988. Responses related to lignification and intravascular periderm formation in carnations resistant to *Fusarium* wilt. Can. J. Bot. 66:784–792.
18 Baayen, R. P., and van der Plas, C. H. 1992. Localization ability, latent period and wilting rate in eleven carnation cultivars with partial resistance to *Fusarium* wilt. Euphytica 59:165–174.
19 Baayen, R. P., Van Eijk, C., and Elgersma, D. M. 1989. Histology of resistant and susceptible carnation cultivars from soil infested with *Fusarium oxysporum* f. sp. *dianthi*. Neth. J. Plant Pathol. 95:3–13.
20 Bacon, C.W., and DeBattista, J. 1991. Endophytic fungi of grasses. Pages 231–256 in: Handbook of Applied Mycology, Soil and Plants, Vol. 1. D. K. Arora, B. Rai, K. G. Mukerji and G. R. Knudsen, eds. Marcel Dekker Inc., New York.
21 Bajar, A., Podila, G. K., and Kolattukudy, P. E. 1991. Identification of fungal cutinase promoter that is inducible by plant signal via a phosphorylated transacting factor. Proc. Natl. Acad. Sci. USA 88:8208–8212.
22 Balasubramanian, R., and Manocha, M.S. 1986. Proteinase, chitinase, and chitosanase activities in germinating spores of *Piptocephalis virginiana*. Mycologia 78:157–163.
23 Balasubramanian, R., and Manocha, M.S. 1992. Cytosolic and membrane bound chitinases of two mucoraceous fungi: a comparative study. Can. J. Microbiol. 38:331–338.
24 Balatinecz, J. J., and Kennedy, R. V. 1967. Maturation of ray parenchyma cells in pine. For. Products J. 17(10):57–64.
25 Bao, W., and Renganathan, V. 1992. Cellobiose oxidase of *Phanerochaete chrysosporium* enhances crystalline cellulose degradation by cellulases. FEBS 302:77–80.
26 Barr, D. P., Shah, M. M., Grover, T. A., and Aust, S. D. 1992. Production of hydroxyl radical by lignin peroxidase from *Phanerochaete chrysosporium*. Arch. Biochem. Biophys. 298:480–485.
27 Barrett-Bee, K., and Hamilton, M. 1984. The detection and analysis of chitinase activity from the yeast form of *Candida albicans*. J. Gen. Microbiol. 130:1857–1861.
28 Bartnicki–Garcia, S. 1973. Fundamental aspects of hyphal morphogenesis. Pages 245–267 in: Microbial Differentiation. J. M. Ashworth and J. E. Smith, eds. Cambridge University Press, London.
29 Bendayan, M. 1981. Ultrastructural localization of nucleic acids by means of nuclease-gold complexes. J. Histochem. Cytochem. 29:531–541.
30 Bendayan, M. 1984. Enzyme-gold electron microscopic cytochemistry: A new affinity technique for the ultrastructural localization of macromolecules. J. Electron Microsc. Tech. 1:349–372.
31 Bendayan, M. 1989. Protein A-gold and protein G-gold postembedding immunoelectron microscopy. Pages 34–95 in: Colloidal Gold. Principles, Methods, and Applications. Vol. 1. M. A. Hayat, ed. Academic Press, New York.
32 Bendayan, M., Nanci, A., and Kan, W. K. 1987. Effect of tissue processing on colloidal cytochemistry. J. Histochem. Cytochem. 35:983–986.
33 Bendayan, M., and Zollinger, M. 1983. Ultrastructural localization of antigenic sites on osmium-fixed tissues applying the proteinA-gold technique. J. Histochem. Cytochem. 31:101–109.
34 Benhamou, N. 1989. Preparation and application of lectin-gold complexes. Pages 96–145 in: Colloidal Gold. Principles, Methods, and Applications. Vol. 1. M. A. Hayat, ed. Academic Press, New York.

35 Benhamou, N. 1989. Ultrastructural study of galacturonic acid in some pathogenic fungi using gold complexed *Aplysia depilans* lectin. Can. J. Microbiol. 35:349–358.
36 Benhamou, N. 1991. Cell surface interactions between tomato and *Clavibacter michiganense* subsp. *michiganense*: localization of some polysaccharides and hydroxyproline-rich glycoproteins in infected host leaf tissues. Physiol. Mol. Plant Pathol. 38:15–38.
37 Benhamou, N. 1992. Ultrastructural detection of β-1,3-glucans in tobacco root tissues infected by *Phytophthora parasitica* var. *nicotianae* using gold-complexed tobacco β-1,3-glucanase. Physiol. Mol. Plant Pathol. 41:351–370.
38 Benhamou, N., Chamberland, H., and Pauzé, F. J. 1989. Implication of pectic components in cell surface interactions between tomato root cells and *Fusarium oxysporum* f. sp. *radicis-lycopersici*. Plant Physiol. 92:995–1003.
39 Benhamou, N., Chamberland, H., Ouellette, G. B., and Pauzé, F. J. 1987. Ultrastructural localization of β–(1–4)–D–glucans in two pathogenic fungi and in their host tissues by means of an exoglucanase–gold complex. Can. J. Microbiol. 33:405–417.
40 Benhamou N., Chamberland, H., Ouellette, G. B., and Pauzé, F. J. 1988. Detection of galactose in two fungi causing wilt disease and in their plant host tissues by means of gold-complexed *Ricinus communis* agglutinin J. Phys. Mol. Plant Pathol. 32:249–266.
41 Benhamou, N., Gilboa-Garber, N., Trudel, J., and Asselin, A. 1988. A new lectin-gold complex for ultrastructural localization of galacturonic acids. J. Histochem. Cytochem. 36:1403–1211.
42 Benhamou, N., Jootsen, M. A. H. J., and de Wit, P. J. G. M. 1990. Subcellular localization of chitinases and of its potential substrate in tomato root tissues infected by *Fusarium oxysporum* f. sp. *radicis-lycopersici*. Plant Physiol. 92:1108–1120.
43 Benhamou, N., Lafontaine, J. G., Joly, J. R., and Ouellette, G. B. 1985. Ultrastructural localization of a toxic glycopeptide produced by *Ophiostoma ulmi*, using monoclonal antibodies. Can. J. Bot. 63:1185–1195.
44 Benhamou, N., Mazau, D., and Esquerré-Tugayé, M. T. 1989. Immunological localization of hydroxyproline-rich glycoproteins in tomato root cells infected by *Fusarium oxysporum* f. sp. *radicis-lycopersici*: Study of a compatible interaction. Phytopathol. 80:163–173.
45 Benhamou, N., and Ouellette, G. B. 1986. Use of pectinases complexed to colloidal gold for the ultrastructural localization of polygalacturonic acids in the cell walls of the fungus *Ascocalyx abietina*. Histochem. J. 18:95–104.
46 Benhamou, N., and Ouellette, G. B. 1986. Ultrastructural characterization of an extracellular sheath on cells of *Ascocalyx abietina*, the scleroderris canker agent of conifers. Can. J. Bot. 65:154–167.
47 Benhamou, N., Ouellette, G. B., Gardiner, R. B., and Day, A. W. 1986. Immunocytochemical localization of antigen binding sites in the cell surface of two ascomycete fungi using antibodies produced against fimbriae from *Ustilago violacea* and *Rhodotorula rubra*. Can. J. Microbiol. 32:871–883.
48 Benyagoub, M., Chamberland, H., and Jabaji-Hare, S. H. 1992. Cytochemical study of the mycoparasitic interaction of *Stachybotrys elegans* with *Rhizoctonia solani*. (Abstr.). Phytopathol. 82:1119.

49 Berg, R.H., Erdos, G.W., Gritzali, M., and Brown, R.D. Jr. 1988. Enzyme–gold affinity labelling of cellulose. J. Electron Microsc. Tech. 8:371–379.
50 Bernards, M. A., and Lewis, N. G. 1992. Alkyl ferulates in wound–healing potato tubers. Phytochem.31:3409–3412.
51 Bes B., Petterson, B., Lennholm, H., Iversen, T., and Eriksson K.-E. 1987. Synthesis, structure and enzymatic degradation of an extracellular glucan produced in nitrogen-starved cultures of the white-rot fungus *Phanerochaete chrysosporium*. Biotechnol. Appl. Biochem. 9:310–318.
52 Biely, P. 1985. Microbial xylanolytic systems. Trends Biotechnol. 3:268–290.
53 Biggs, A. R. 1984. Intracellular suberin: occurrence and detection in tree bark. IAWA Bull. 5:243–248.
54 Biggs, A. R. 1986. Comparative anatomy and host response of two peach cultivars inoculated with *Leucostoma cincta* and *L. persoonii*. Phytopathol. 76:905–912.
55 Biggs, A. R. 1987. Occurrence and location of suberin in wound reaction zones in xylem of 17 tree species. Phytopathol. 77:718–725.
56 Biggs, A. R. 1992. Anatomical and physiological responses of bark tissues to mechanical injury. Pages 13–40 in: Defense Mechanisms of Woody Plants against Fungi. R. A. Blanchette and A. R. Biggs, eds. Springer–Verlag, Berlin.
57 Biggs, A. R. 1992. Responses of angiosperm bark tissues to fungi causing cankers and canker rots. Pages 41–61 in: Defense Mechanisms of Woody Plants against Fungi. R. A. Blanchette and A. R. Biggs, eds. Springer–Verlag, Berlin.
58 Biggs, A. R., and Alm, G. 1992. Response of peach bark tissues to inoculation with epiphytic fungi alone and in combination with *Leucostoma cincta*. Can. J. Bot. 70:186–191.
59 Biggs, A. R., Miles, N. W., and Bell, R. L. 1992. Heritability of suberin accumulation in wounded peach bark. Phytopathol. 82:83–86.
60 Blaiseau, P.-L., Kunz, C., Grison, R., Bertheau, Y., and Brygoo, Y. 1992. Cloning and expression of a chitinase gene from the hyperparasitic fungus *Aphanocladium album*. Curr. Genet. 21:61–66.
61 Blaiseau, P.-L., Lafay, J–F. 1992. Primary structure of a chitinase–encoding gene (chi 1) from the filamentous fungus *Aphanocladium album*: similarities to bacterial chitinases. Gene 120:243–248.
62 Blanchette, R. A. 1991. Delignification by wood-decay fungi. Annu. Rev. Phytopathol. 29:381–398.
63 Blanchette, R. A. 1992. Anatomical responses of xylem to injury and invasion by fungi. Pages 76–95 in: Defense Mechanisms of Woody Plants against Fungi. R. A. Blanchette, and A. R. Biggs, eds. Springer–Verlag, Berlin.
64 Blanchette, R. A., Abad, A. R., Cease, K. R., Lovrien, R. E., and Leathers, T. D. 1989. Colloidal gold cytochemistry of endo-1,4-β-glucanase, 1,4-β-D-glucan cellobiohydrolase, and endo 1,4-β-xylanase: Ultrastructure of sound and decayed birch wood. Appl. Environ. Microbiol. 55:2293–2301.
65 Blanchette, R. A., Abad, A. R., Farrell, R. L., and Leathers, R. L. 1989. Detection of lignin peroxidase and xylanase by immunocytochemical labeling in wood decayed by basidiomycetes. Appl. Environ. Microbiol. 55:1457–1465.
66 Blanchette, R. A., Akhtar, M., and Attridge, M. C. 1992. Using simons

stain to evaluate fiber characteristics of biomechanical pulps. Tappi J. 75:121–124.

67 Blanchette, R. A., Burnes, T. A., Eerdmans, M. M., and Akhtar, M. 1992. Evaluating isolates of *Phanerochaete chrysosporium* and *Ceriporiopsis subvermispora* for use in biological pulping processes. Holzforschung 46:109–115.

68 Blanchette, R. A., Obst, J. R., Hedges, J. I., and Weliky, K. 1988. Resistance of hardwood vessels to degradation by white rot basidiomycetes. Can. J. Bot. 66:1841–1847.

69 Blanchette, R. A., Obst, J. R., and Timell, T. E. 1994. Biodegradation of compression wood and tension wood by white and brown rot fungi. Holzforschung. In press.

70 Blanchette, R. A., Otjen, L., and Carlson, M. C. 1987. Lignin distribution in cell walls of birch wood decayed by white rot basidiomycetes. Phytopathol. 77:684–690.

71 Bland, D. E., Foster, R. C., and Logan, A. F. 1971. The mechanism of permanganate and osmium tetroxide fixation and the distribution of lignin in the cell walls of *Pinus radiata*. Holzforschung 25:137–142.

72 Boddy, L. 1992. Microenvironmental aspects of xylem defenses to wood decay fungi. Pages 96–132 in: Defense Mechanisms of Woody Plants against Fungi. R. A. Blanchette and A. R. Biggs, eds. Springer–Verlag, Berlin.

73 Bonfante, P. 1988. The role of the cell–wall as a signal in mycorrhizal associations. Pages 219–235 in: Cell to Cell signals in Plant, Animal, and Microbial Symbiosis. S. Scannerini, D. C. Smith, P. Bonfante and V. Gianinazzi–Pearson, eds. Springer Verlag, Berlin.

74 Bonfante, P., and Perotto, S. 1992. Plants and endomycorrhizal fungi: the cellular and molecular basis of their interaction. Pages 445–470 in: Molecular Signals in Plant–Microbe Communications. D. P. S. Verma, ed. CRC Press, Boca Raton, Ann Arbor.

75 Bonfante, P., Perotto, S., Testa, B., and Faccio, A. 1987. Ultrastructural localization of cell surface sugar residues in ericoid mycorrhizal fungi by gold-labeled lectins. Protoplasma 139:25–35.

76 Bonfante, P., and Scannerini, S. 1992. The cellular basis of plant–fungus interchanges in mycorrhizal association. Pages 65–101 in: Mycorrhizal Functioning. M. F. Allen, ed. Chapman and Hall, New York London.

77 Bonfante, P., Vian, B., Perotto, S., Faccio, A., and Knox, J. P. 1990. Cellulose and pectin localization in roots of mycorrhizal *Allium porrum*: labeling continuity between host cell wall and interfacial material. Planta 180:537–547.

78 Bostock, R. M., and Middleton, G. E. 1987. Relationship of wound periderm formation to resistance to *Ceratocystis fimbriata* in almond bark. Phytopathol. 77:1174–1180.

79 Bourson, C., Favey, S., Reverchon, S., and Robert-Baudouy, J. 1993. Regulation of the expression of a pel*A* fusion in *Erwinia chrysanthemi* and demonstration of the synergistic action of plant extract with polygalacturonate on pectate lyase synthesis. J. Gen. Microbiol. 139:1–9.

80 Bracker, C. E., and Littlefield, L. J. 1973. Structural concepts of host–pathogen interfaces. Pages 159–317 in: Fungal Pathogenicity and the Plant's Response. R. J. W. Byrde and C. V. Cutting, eds. Academic Press, London.

81 Bradley, B. J., Kjelborn, P., and Lamb, C. J. 1992. Elicitor and wound-induced oxidative cross-linking of a proline-rich plant cell wall protein: a

novel, rapid defense response. Cell 70:21-30.

82 Brawerman, G. 1989. mRNA decay, finding the right targets. Cell 57:9-10.

83 Brewin, N. J., Robertson, J. G., Wood, E. A., Wells, B., Larkins, A. P., Galfre, G., and Bucher, G. W. 1985. Monoclonal antibodies to antigens in the peribacteroid membrane for Rhizobium-induced root nodules of pea cross react with plasma membranes and Golgi bodies. EMBO J. 4:605–611.

84 Brisson, J. D., Robb, J., and Peterson, R. L. 1976. Phenolic localization by ferric chloride and other iron compounds. Microscop. Soc. Canada 3:174–176.

85 Broglie, K., Chet, I., Holliday, M., Cressman, R., Biddle, P., Knowlton, S., Mauvais, C. J., and Broglie, R. 1991. Transgenic plants with enhanced resistance to the fungal pathogen *Rhizoctonia solani*. Science 254:1194–1197

86 Brundrett, M., and Kendrick, B. 1990. The roots and mycorrhizas of herbaceous woodland plants.II.Structural aspects of morphology. New Phytol. 114:469–479.

87 Buchala, A. J., and Leisola, M. 1987. Structure of the β-D-glucan secreted by *Phanerochaete chrysosporium* in continuous culture. Carbohydr. Res. 175:146–149.

88 Bullock, S., Ashford, A. E., and Willetts, H. J. 1980. The structure and histochemistry of sclerotia of *Sclerotinia minor* Jagger. II. Histochemistry of extracellular substances and cytoplasmic reserves. Protoplasma 104:333–351.

89 Bushnell, W. R. 1972. Physiology of fungal haustoria. Annu. Rev. Phytopathol. 10:151–176.

90 Cabib, E., Silverman, S.J., and Shaw, J.A. 1992. Chitinase and chitin synthase 1: counterbalancing activities in cell separation of *Saccharomyces cerevisiae*. J. Gen. Microbiol. 138:97–102.

91 Cabral, D., Stone, J. K., and Carroll, G. C. 1993. The internal mycobiota of *Juncus* spp.: microscopic and cultural observations of infection patterns. Mycol. Res. 97:367–376.

92 Calvete, J. S. 1992. Function of cutinolytic enzymes in the infection of gerbera flowers by *Botrytis cinerea* Thesis, University of Utrecht, Netherlands, Pages 75–84.

93 Carpita, N. C., and Gibeaut, D. M. 1993. Structural models of primary cell walls in flowering plants: consistency of molecular structure with the physical properties of the walls during growth. Plant J. 3:1–30.

94 Carroll, G. C. 1988. Fungal endophytes of stems and leaves: from latent pathogen to mutualistic symbiont. Ecology 69:2–9.

95 Carroll, G. C., and Carroll, F. E. 1978. Studies on the incidence of coniferous needle endophytes in the Pacific Northwest. Can. J. Bot. 56:3034–3043.

96 Casagrande, F., and Ouellette, G. B. 1971. A technique to study the development in wood of soft rot fungi and its application to *Ceratocystis ulmi*. Can. J. Bot. 49:155–159.

97 Cervone, S., Castoria, R., Spanu, P., and Bonfante, P. 1988. Pectinolytic activity in some ericoid mycorrhizal fungi. Trans. Br. mycol. Soc. 91:537–539.

98 Chamberland, H., Benhamou, N., Ouellette, G. B., and Pauzé, F. J. 1989. Cytochemical detection of saccharide residues in paramural bodies formed in tomato root cells infected by *Fusarium oxysporum* f. sp. *radicis-lyco-*

persici. Physiol. Mol. Plant Pathol. 34:131–146.

99 Chamberland, H., Charest, P. M., Ouellette, G. B., and Pauzé, F. J. 1985. Chitinase-gold complex used to localize chitin ultrastructurally in tomato root cells infected by *Fusarium oxysporum* f. sp. *radicis-lycopersici*, compared with a chitin specific gold-conjugated lectin. Histochem. J. 17:313–320.

100 Chamberland, H., Ouellette, G. B., Pauzé, F. J., and Charest, P. M. 1991. Immunocytochemical localization of tomato pectin esterase in root cells of tomato plants infected by *Fusarium oxysporum* f. sp. *radicis-lycopersici.* Can. J. Bot. 69:1265–1274.

101 Chang, H–M., and Allan, G. G. 1971. Oxidation. Pages 433–483 in: Lignins: Occurrence, Formation, Structure and Reactions. K. V. Sarkanen, and C. H. Ludwig, eds. Wiley–Interscience, NY.

102 Chanzy, H. 1990. Aspects of cellulose structure. Pages 3–12 in: Cellulose Sources and Exploitation. Industrial Utilization, Biotechnology and Physico–chemical Properties. J. F. Kennedy, G. O. Phillips and P. A. Williams, eds. Ellis Horwood.

103 Chanzy, H., Grosrenaud, A., Joseleau, J. P., Dube, M., and Marchessault, R. H. 1982. Crystallization behaviour of glucomannan. Biopolymers 21:301–319.

104 Chapela, I. H., Petrini, O., and Hagmann, L. 1991. Monolignol glucosides as specific recognition messengers in fungus–plant symbioses. Physiol. Mol. Plant Pathol. 39:289–298.

105 Charest, P. M., Ouellette, G. B., and Pauzé, F. J. 1984. Cytological observations of early infection process of *Fusarium oxysporum* f. sp. *radicis-lycopersici.* Can. J. Bot. 62:1232–1244.

106 Chaubal, R., Wilmot, V. A., and Wynn, W. K. 1991. Visualization, adhesiveness, and cytochemistry of the extracellular matrix produced by urediniospore germ tubes of *Puccinia sorghi.* Can. J. Bot. 69:2044–2054.

107 Cheong, J. J., Birberg, W., Fügedi, P., Pilotti, A, Garegg, P. J., Hong, N., Ogawa, T., and Hahn, M. G., 1991. Structure activity relationships of oligo-β-glucoside elicitors of phytoalexin accumulation in Soybean. Plant Cell 3:127–136.

108 Cheong, J.-J., and Hahn, M. G. 1991. A specific, high-affinity binding of a synthetic heptaglucoside and fungal glucan phytoalexin elicitor to soybean membranes. Plant Cell 3:137-147.

109 Cherif, M., and Benhamou, N. 1990. Cytochemical aspects of chitin breakdown during the parasitic action of a *Trichoderma* sp. on *Fusarium oxysporum* f. sp. *radicis–lycopersici.* Phytopathol. 80:1406–1414.

110 Chet, I. 1987. *Trichoderma* – application, mode of action, and potential as a biocontrol agent of soil borne plant pathogenic fungi. Pages 137–160 in: Innovative Approaches to Plant Disease Control. I. Chet., ed. John Wiley, New York.

111 Chet, I., Ordentlich, A., Shapira, R., and Oppenheim, A. 1990. Mechanisms of biocontrol of soil–borne plant pathogens by Rhizobacteria. Plant and Soil 129:85–92.

112 Chong, J., and Harder, D. E. 1980. Ultrastructure of haustorium development in *Puccinia coronata avenae* I. Cytochemistry and electron probe X-ray analysis of the haustorial neck ring. Can. J. Bot. 58:2496–2505.

113 Chong, J., Harder, D. E., and Rohringer, R. 1986. Cytochemical studies on *Puccinia graminis* f.sp. *tritici* in a compatible wheat host. II. Haustorium mother cell walls at the host cell penetration site, haustorial walls, and the extrahaustorial matrix. Can. J. Bot. 64:2561–2575.

114 Clark, E. M., White, J. F., and Patterson, R. M. 1983. Improved histochemical techniques for the detection of *Acremonium coenophialum* in the tall grass fescue and methods of *in vitro* culture of the fungus. J. Microbiol. Meth. 1:149–155.
115 Clay, K. 1988. Fungal endophytes of grasses: a defensive mutualism between plants and fungi. Ecology 69:10–16.
116 Clay, K. 1990. Fungal endophytes of grasses. Annu. Rev. Ecol. Syst. 21:275–297.
117 Coffey, M. D. 1983. Cytochemical specialization at the haustorial interface of a biotrophic fungal parasite, *Albugo candida*. Can. J. Bot. 61:2004–2014.
118 Collmer, A., Bauer, D. W., He, S. Y., Lindeberg, M., Kelmer, S., Rodrigues–Balenzuela, P., Burr, T. J., and Chatterjee, A. K. 1991. Pectic enzyme production and bacterial plant pathogenicity. Pages 65–72 in: Advances in Molecular Genetics of Plant–Microbe Interactions, Vol. 1. H. Hennecke and D. P. S. Verma, eds. Kluwer Academic Publishers, The Netherlands.
119 Collmer, A., and Keen, N. T. 1986. The role of pectic enzymes in plant pathogenesis. Annu. Rev. Phytopathol. 24:383–409.
120 Cooke, R.C. 1977. The biology of symbiotic fungi. Wiley, London.
121 Cooper, R. M., Wardman, P. A., and Skelton, J. E. M. 1981. The influence of cell walls from host and non–host plants on the production and activity of polygalacturonide–degrading enzymes from fungal pathogens. Physiol. Plant Pathol. 18:239–255.
122 Cor, F. B., Witteveen, B., Vennhuis, M., and Visser, R. 1992. Localization of glucose oxidase and catalase activities in *Aspergillus niger*. Appl. Environ. Microbiol. 58:1190–1194.
123 Corbett, N.H. 1965. Micro-morphological studies on the degradation of lignified cell walls by ascomycetes and Fungi Imperfecti. Inst. Wood Sci. J. 3:18–29.
124 Corbin, D. R., Sauer, N., and Lamb, C. J. 1987. Differential regulation of a hydroxyproline-rich glycoprotein gene family in wounded and infected plants. Mol. Cell. Biol. 7:4337–4344.
125 Crawford, M. S., and Kolattukudy, P. E. 1987. Pectate lyase from *Fusarium solani* f.sp. *pisi*, Purification, characterization, in vitro translation of the mRNA, and involvement in pathogenicity. Arch. Biochem. Biophys. 258:196–205.
126 Daniel, G., Nilsson, T., and Pettersson, B. 1989. Intra- and extracellular localization of lignin peroxidase during the degradation of solid wood and wood fragments by *Phanerochaete chrysosporium* by using transmission electron microscopy and immuno-gold labeling. Appl. Environ. Microbiol. 55:871–881.
127 Daniel, G., Petterson, B., Nilsson, T., and Volc, J. 1990. Use of immunogold cytochemistry to detect Mn(II)-dependant and lignin peroxidases in wood degraded by the white-rot fungi *Phanerochaete chrysosporium* and *Lentinula edodes*. Can. J. Bot. 68:920–933.
128 Daniel, G., Volc, J., Kubatova, E., and Nilsson, T. 1992. Ultrastructural and Immunocytochemical Studies on the H_2O_2-producing enzyme pyranose oxidase in *Phanerochaete chrysosporium* grown under liquid culture conditions. Appl. Environ. Microbiol. 58:3667–3676.
129 Dantzig, A. H., Zuckerman, S. H., and Andonov–Roland, M. M. 1986. Isolation of a *Fusarium solani* mutant reduced in cutinase activity and virulence. J. Bacteriol. 168:911–916.

130 Day, A. W., and Poon, N. H. 1975. Fungal fimbriae. I. Their role in conjugation in *Ustilago violacea*. Can. J. Microbiol. 21:547–557.
131 De La Cruz, J., Hidalgogallego, A., Lora, J. M., Benitez, T., Pintortoro, J. A., and Leobell, A. 1992. Isolation and characterization of 3 chitinases from *Trichoderma harzianum*. Eur. J. Biochem. 206:859–867.
132 Dea, I. C. M., and Rees, D. A. 1973. Aggregation with change of conformation in solutions of hemicellulose xylans. Carbohydr. Res. 29:363–372.
133 DeMey, J., and Moeremans, M. 1986. Raising and testing polyclonal antibodies for immunocytochemistry. Pages 3–12 in: Immunocytochemistry: Modern Methods and Applications. J. M. Polak and S. Van Noorden, eds. John Wright and Sons, Bristol.
134 Dence, C. W. 1992. The determination of lignin. Pages 33–61 in: Methods in Lignin Chemistry. S. Y. Lin and C. W. Dence, eds. Springer–Verlag. Berlin–Heidelberg.
135 Dexheimer, J., and Pargney, J. C. 1991 Comparative anatomy of the host–fungus interface in mycorrhizas. Experientia 47:312–321.
136 Dickerson, A. G., and Baker, R. C. F. 1979. The binding of enzymes to fungal β-glucans. J. Gen. Microbiol. 112:67–75.
137 Dickinson, K., Keer, V., Hitchcock, C. A., and Adams, D. J. 1991. Microsomal chitinase activity from *Candida albicans*. Biochim. Biophys. Acta 1073:177–182.
138 Dickman, M. B., and Patil, S. S. 1986. Cutinase deficient mutants of *Colletotrichum gloeosporioides* are nonpathogenic to papaya fruit. Physiol. Mol. Plant Pathol. 28:235–242.
139 Dickman, M.B., Patil, S.S., and Kolattukudy, P.E. 1982. Purification, characterization and role in infection of an extracellular cutinolytic enzyme from *Colletotrichum gloeosporioides* Penz. on *Carica papaya* L. Physiol. Plant Pathol. 20:333–347.
140 Dickman, M. B., Podila, G. K., and Kolattukudy, P. E. 1989. Insertion of cutinase gene into a wound pathogen enables it to infect intact host. Nature 342:446–448.
141 Din, N., Gilkes, N. R., Tekant, B., Miller, R. C., Jr., Warren, A. J., and Kilburn, D. G. 1991. Non–hydrolytic disruption of cellulose fibres by the binding domain of a bacterial cellulase. Biotechnol. 9:1096–1099.
142 Donaldson, L. A. 1985. Critical assessment of interference microscopy as a technique for measuring lignin distribution in cell walls. New Zeal. J. For. Sci. 15:349–360.
143 Donaldson, L. A. 1992. Lignin distribution during latewood formation in *Pinus radiata* D. Don. IAWA Bull. 13:381–387.
144 Dubourdieu, D., and Ribéreau-Gayon, P. 1981. Structure of the extracellular β-D-glucan from *Botrytis cinerea*. Carbohydr. Res. 93:294–299.
145 Duddridge, J.A. 1987. Specificity and recognition in ectomycorrhizal associations. Pages 159–317 in: Fungal Infection of Plants. G. F. Pegg and P. G. Ayres, eds. Cambridge University Press, Cambridge.
146 Dumas–Gaudot, E., Grenier, J., Furlan, V., and Asselin, A. 1992. Chitinase, chitosanase and ß–1,3 glucanase activities in *Allium* and *Pisum* roots colonized by *Glomus* species. Plant Science 84:17–24.
147 Dutton, G. G. S., and Joseleau, J. P. 1977. Hemicelluloses of Red wood (*Sequoia sempervirens*). Cell. Chem. Technol. 11:313–319.
148 Effland, M. 1977. Modified procedure to determine acid–insoluble lignin in wood and pulp. Tappi 60:143–144.
149 Ehrlich, M. A., and Ehrlich, H. G. 1971. Fine structure of the host-para-

site interfaces in mycoparasitism. Annu. Rev. Phytopathol. 9:155–184.
150 Elad, Y., Barak, R., and Chet, I. 1982. Degradation of plant pathogenic fungi by *Trichoderma harzianum*. Can. J. Microbiol. 18:719–725.
151 Elango, N., Correa, J. U., and Cabib, E. 1982. Secretory character of yeast chitinase. J. Biol. Chem. 257:1398–1400.
152 Elgersma, D. M. 1970. Length and diameter of xylem vessels as factors in resistance of elms to *Ceratocystis ulmi*. Neth. J. Plant Pathol. 76:179–182.
153 Elgersma, D. M., and Miller, H. J. 1977. Tylose formation in elms after inoculation with an aggressive or a non–aggressive strain of *Ophiostoma ulmi* or with a non–pathogen to elms. Neth. J. Plant Pathol. 83:241–243.
154 Emmett, R. W., and Parbery, D. G. 1975. Appressoria. Annu. Rev. Phytopathol. 58:147–167.
155 Enoki, A., Tanaka, H., and Fuse, G. 1989. Relationship between degradation of wood and production of H_2O_2–producing one–electron oxidases by brown–rot fungi. Wood Sci. Technol. 23:1–12.
156 Epstein, L., Laccetti, L. B., and Staples, R. C. 1987. Cell-substratum adhesive protein involved in surface contact responses of the bean rust fungus. Physiol. Mol. Plant Pathol. 30:373–388.
157 Eriksson, K. E. L., Blanchette, R. A., and Ander, P. 1990. Microbial and Enzymatic Degradation of Wood and Wood Components. Springer-Verlag, Berlin, Heidelberg, etc.
158 Espejo, E., and Agosin, E. 1991. Production of oxalic acid by brown rot fungi. Appl. Environ. Microbiol. 57:1980–1986.
159 Espelie, K. E., Franceschi, V. R., and Kolattukudy, P. E. 1986. Immunocytochemical localization and time–course of appearance of an anionic peroxidase associated with suberization in wound–healing potato tuber tissue. Plant Physiol. 81:487–492.
160 Espinosa-Garcia, F. J., and Langenheim, J.H. 1990. The endophytic fungal community in leaves of a coastal redwood population – diversity and spatial patterns. New Phytol. 116:89–97.
161 Evans, C. S., Gallagher, I. M., Atkey, P. T., and Wood, D. A. 1991. Localisation of degradative enzymes in white-rot decay of lignocellulose. Biodegradation 2:93–106.
162 Evans, R. C., Stempen H., and Stewart, S. J. 1981. Development of hyphal sheaths in *Bipolaris maydis* race T. Can. J. Bot. 59:453–459.
163 Faix, O., Mozuch, M. D., and Kirk, T. K. 1985. Degradation of gymnosperm (guaiacyl) vs. angiosperm (syringyl/guaiacyl) lignins by *Phanerochaete chrysosporium*. Holzforschung 39:203–208.
164 Fengel, D., and Wegener, G. 1984. Wood chemistry, Ultrastructure, Reactions. W. De Gruyter.
165 Fergus, B. J., Procter, A. R., Scott, J. A. N., and Goring, D. A. I. 1969. The distribution of lignin in spruce wood as determined by UV microscopy. Wood Sci. Technol. 3:117–138.
166 Fernandez, M. R., and Heath, M. C. 1989. Interactions of the nonhost French bean plant (*Phaseolus vulgaris*) with parasitic and saprophitic fungi. III Cytologically-detectable responses. Can. J. Bot. 67:676–686.
167 Fieschi, M., Alloatti G., Sacco, S., and Berta, G. 1992. Membrane potential hyperpolarisation in vesicular arbuscular mycorrhizae of *Allium porrum* L.: a non–nutritional long–distance effect of the fungus. Protoplasma 168:136–140.
168 Flournoy, D. S., Kirk, T. K., and Highley, T. L. 1991. Wood decay by brown–rot fungi: changes in pore structure and cell wall volume.

Holzforschung 45:383–388.
169 Foisner, R., Messner, K., Stachelberger, H., and Röhr, M. 1985. Wood decay by basidiomycetes: extracellular tripartite membranous structures. Trans. Br. mycol. Soc. 85:257–266.
170 Forrester, I. T., Grabski, A. C., Burgess, R. R., and Leatham, G. F. 1988. Manganese, Mn–dependent peroxidases, and the biodegradation of lignin. Biochem. Biophys. Res. Commun. 157:992–999.
171 Freudenberg, K., and Harkin, J.M. 1968. The glucosides of cambial sap of spruce. Phytochem. 2:189–193.
172 Fukazawa, K., and Imagawa, H. 1981. Quantitative analysis of lignin using an UV microscopic image analyzer. Variation within one growth increment. Wood Sci. Technol. 15:45–55.
173 Galfre, G., and Milstein, C. 1981. Preparation of monoclonal antibodies: strategies and procedures. Meth. Enzym. 73B:3–46.
174 Gall, J. G., and Pardue, M. L. 1969. Formation and detection of RNA-DNA hybrid molecules in cytological preparations. Proc. Natl. Acad. Sci. USA 63:378-383.
175 Gallagher, M. I., and Evans, C. S. 1990. Immunogold cytochemical labelling of β-glucosidase in the white-rot fungus *Coriolus versicolor*. Appl. Microbiol. Biotechnol. 32:588–593.
176 Galliano, H., Gas, G., and Boudet, A. M. 1991. Lignin degradation by *Rigidoporus lignosus* involves synergistic action of two oxidizing enzymes–Mn peroxidase and laccase. Enzyme Microb. Technol. 13:478–482.
177 Garbow, J. R., Ferrantello, L. M., and Stark, R. E. 1989. ^{13}C Nuclear magnetic resonance study of suberized potato cell wall. Plant Physiol. 90:783–787.
178 Garcia Romera, I., Garcia–Garrido, J. M., Martinez–Molina, E., and Ocampo, J. A. 1990. Possible influence of hydrolytic enzymes on vesicular–arbuscular mycorrhizal infection of alfalfa. Soil Biol. Biochem. 22:149–152.
179 Gardiner, R. B., and Day, A. W. 1988. Surface proteinaceous fibrils (fimbriae) on filamentous fungi. Can. J. Bot. 66:2474–2484.
180 Garriock, M. L., Peterson, R. L., and Ackerley, C. A. 1989. Early stages in colonization of *Allium porrum* (leek) roots by the VAM fungus, *Glomus versiforme*. New Phytol. 112:85–94.
181 Gay, J. L., Salzberg, A., and Woods, A. M. 1987. Dynamic experimental evidences for the plasma membrane ATPase domain hypothesis of haustorial transport for ionic coupling of the haustorium of *Erysiphe graminis* to the host cell (*Hordeum vulgare*). New Phytol. 107:541–548.
182 Geiger, J. P., Huguenin, B., Nicole, M., and Nandris, D. 1986. Laccases of *Rigidoporus lignosus* and *Phellinus noxius*. II – Effects of *R. lignosus* laccase L1 on thioglycolic lignin of *Hevea*. Appl. Biochem. Biotechnol. 13:97–111.
183 Gianinazzi–Pearson, V. Bonfante-Fasolo, P., and Dexheimer, J. 1986. Ultrastructural studies of surface interactions during adhesion and infection by ericoid endomycorrhyzal fungi. Pages 273–282 in: Recognition in Microbe-Plant Symbiotic and Pathogenic Interactions. B. Lugtenberg, ed. NATO ASI Series, Vol. 4.
184 Gianinazzi–Pearson, V., Smith, S. E., Gianinazzi, S., and Smith, F. A. 1991. Enzymatic studies on the metabolism of vesicular–arbuscular mycorrhizas. V. Is H^+–ATPase a component of ATP–hydrolysing enzyme activities in plant–fungus interfaces? New Phytol. 117:61–74.
185 Gilkes, N. R., Henrissat, B., Kilburn, D. G., Miller, R. C., and Warren,

A. J. 1991. Domains in microbial β–1,4–glycanases: sequence conservation, function and enzyme families. Microbiol. Rev. 55:303–315.

186 Giovannetti, M., Avio, L., Sbrana, C., and Citernesi, A. S. 1993. Factors affecting appressorium development in the vesicular–arbuscular mycorrhizal fungus *Glomus mosseae* (Nicol. and Gerd.) Gerd. and Trappe. New Phytol. 123:115–122.

187 Gold, M. H., Wariishi, H., and Valli, K. 1989. Extracellular peroxidases involved in lignin degradation by the white rot basidiomycete *Phanerochaete chrysosporium*. Pages 127–140 in: Biocatalysis in Agricultural Biotechnology. J. R. Whitaker and P. E. Sonwet, eds. American Chemical Society Series 389. Washington, D.C.

188 Goldstein, I. J., Hughes, R. C., Monsigny, M., Osawa, T., and Sharon, N. 1980. What should be called a lectin? Nature 285:66.

189 Gollotte,A., Gianinazzi–Pearson, V., Giovannetti, M., Sbrana, C., Avio, L., and Gianinazzi, S. 1993. Cellular localization and cytochemical probing of resistance reactions to arbuscular mycorrhizal fungi in a 'locus a' myc$^-$ mutant of *Pisum sativum* L. Planta 191:112–122.

190 González–Candelás, L., and Kolattukudy, P. E. 1992. Isolation and analysis of a novel inducible pectate lyase gene from the phytopathogenic fungus *Fusarium solani* f. sp. *pisi* (*Nectria haematococca* mating population VI). J. Bacteriol. 174:6343–6349.

191 Gooday, G. W., Humphreys, A. M., and McIntosh, W. H. 1986. Roles of chitinases in fungal growth. Pages 83–91 in: Chitin in Nature and Technology. R. A. A. Muzzarelli, C. Jeuniaux and G. W. Gooday, eds. Plenum Press, New York.

192 Goodman, S. L. Hodges, G. M. Trejdosiewics, L. K., and Livingston, D. C. 1979. Colloidal gold probes. A further evaluation. Pages 619–628 in: Scanning Electron Microscopy. III SEM Inc, AMF O'Hare, Il.

193 Green III, F., Clausen, C. A., Larsen, M. J., and Highley, T. L. 1992. Immuno-scanning electron microscopic localization of extracellular wood-degrading enzymes within the fibrillar sheath of the brown-rot fungus *Postia placenta*. Can. J. Microbiol. 38:898–904.

194 Green III, F., Larsen, M. J., Hackney, J. M., Clausen, C. A., and Highley, T. L. 1992. Acid-mediated depolymerization of cellulose during incipient brown-rot decay by *Postia placenta*. Pages 267–272 in: Biotechnology in Pulp and Paper Industry. M. Kuwahara and M. Shimada (eds). Uni Publishers Co. Ltd.

195 Greenwood, J. S., Keller, G. A., and Chrispeels, M. J. 1984. Localization of phytohemagglutinin in the embryonic axis of *Phaseolus vulgaris* with ultrathin cryosections embedded in plastic after indirect immunolabeling. Planta 161:548–555.

196 Gross, G. G. 1981. Phenolic Acids. Pages 301–316 in: The Biochemistry of Plants, Vol. 7. P. K. Stumpf, ed. Academic Press, London.

197 Gunaratna, K. R. 1993. Isolation, purification and some properties of extracellular chitinase from *Acremonium obclavatum* an antagonist to *Puccinia arachidis*. Ph.D. thesis, University of Madras, Madras 25, India.

198 Habem T., Shimada, M., and Higuchi, T. 1985. Biomimetic approach to lignin degradation. I. H_2O_2–dependent C–C bond cleavage of the lignin model compounds with a natural iron porphyrin and imidazole complex. Mokuzai Gakkaishi 31:54–55.

199 Hahlbrock, K., and Scheel, D. 1989. Physiology and molecular biology of phenylpropanoid metabolism. Annu. Rev. Plant Physiol. 40:347–369.

200 Hainfeld, J. F., and Furuya, F. R. 1992. A 1.4-nm gold cluster covalently

attached to antibodies improves immunolabeling. J. Histochem. Cytochem. 40:177–184.
201 Hale, M. D., and Eaton, R. A. 1985. The ultrastructure of soft rot fungi. I. Fine hyphae in wood cell walls. Mycologia 77:447–463.
202 Hale, M. D., and Eaton, R. A. 1985. The ultrastructure of soft rot fungi. II. Cavity-forming hyphae in wood cell walls. Mycologia 77:594–605.
203 Hale, M. D., and Eaton, R. A. 1985. Oscillatory growth of fungal hyphae in wood cell walls. Trans. Br. mycol. Soc. 84:277–288.
204 Hall, J. L., and Hawes, C. 1991. Electron microscopy of plant cells. Academic Press, London, New York.
205 Hammel, K. E., and Moen, M. A. 1991. Depolymerization of a synthetic lignin in vitro by lignin peroxidase. Enzyme Microb. Technol. 13:15–18.
206 Hämmerli, U. A., Brändle, U. E., Petrini, O., and McDermott, J. M. 1992. Differentiation of isolates of *Discula umbrinella* (teleomorph: *Apiognomonia errabunda*) from beech, chestnut and oak using RAPD markers. Molec. Plant–Microbe Interact. 5:479–483.
207 Hammerschmidt, R. 1984. Rapid deposition of lignin in potato tuber tissue as a response to fungi non–pathogenic on potato. Physiol. Plant Pathol. 24:33–42.
208 Hammerschmidt, R., Bonnen, A. M., Bergstrom, G. C., and Baker, K. K. 1985. Association of epidermal lignification with nonhost resistance of cucurbits to fungi. Can. J. Bot. 63:2393–2398.
209 Handley, D. A. 1989. Methods for synthesis of colloidal gold. Pages 13–30 in: Colloidal Gold: Principles, Methods, and Applications. Vol.1. M. A. Hayat, ed. Academic Press, New York.
210 Harder, D. E., and Chong J. 1984. Structure and physiology of haustoria. Pages 431–476 in: The Cereal Rusts. Vol. I. W. R. Bushnell and A. P. Roelfs, eds. Academic Press, New York.
211 Harder, D. E., Chong, J., Rohringer, R., Mendgen, K., Schneider, A., Welter, K., and Knauf, G. 1986. Ultrastructure and cytochemistry of extramural substances associated with intercellular hyphae of several rust fungi. Can. J. Bot. 67:2043–2051.
212 Harley, J. L. 1989. The significance of mycorrhiza. Mycol. Res. 92:92–129.
213 Hayat, M. A. 1981. Principles and techniques of electron microscopy. Biological Applications. Vol 1, 3rd edition. University Park Press, Baltimore.
214 Hazlewood, G. P., and Gilbert, H. J. 1992. The molecular architecture of xylanases from *Pseudomonas fluorescens* subsp. *cellulosa*. Pages 259–273 in: Progress in Biotechnology 7. Xylans and Xylanases. J. Visser, E. Beldman, M.A. Kusters-van Someren and A.G.J. Voragen (eds). Elsevier, Amsterdam.
215 Heath, M. C. 1979. Partial characterization of the electron–opaque deposits formed in the non–host plant, French bean, after cowpea rust infection. Physiol. Plant Pathol. 15:141–148.
216 Heath, M. C. 1980. Effects of infection by compatible species or injection of tissue extracts on the susceptibility of nonhost plants to rust fungi. Phytopathol. 70:356–360.
217 Heath, M. C. 1980. Reactions of nonsuscepts to fungal pathogens. Annu. Rev. Phytopathol. 18:211–236.
218 Heath, M. C. 1981. Nonhost resistance. Pages 201–217 in: Plant Disease Control: Resistance and Susceptibility. R. C. Staples and C. H. Toenniessen, eds. John Wiley and Sons, New York.

219 Heath, M. C. 1981. The suppression of the development of silicon–containing deposits in French bean leaves by exudates of the bean rust fungus and extracts from bean rust–infected tissue. Physiol. Plant Pathol. 18:149–155.

220 Heath, M. C. 1987. Evolution of plant resistance and susceptibility to fungal invaders. Can. J. Plant Pathol. 9:389–397.

221 Heath, M. C., and Heath, I. B. 1971. Ultrastructure of an immune and a susceptible reaction of cowpea leaves to rust infection. Physiol. Plant Pathol. 1:277–287.

222 Hedges, J. I., Blanchette, R. A., Weliky, K., and Devol, A. H. 1988. Effects of fungal degradation on the CuO oxidation products of lignin: A controlled laboratory study. Geochim. Cosmochim. Acta 52:2717–2726.

223 Herman, E. M. 1989. Colloidal gold labeling of acrylic resin-embedded plant tissues. Pages 304–321 in: Colloidal Gold: Principles, Methods, and Applications. Vol. 2. M. A. Hayat, ed. Academic Press, New York.

224 Higuchi, T. 1986. Catabolic pathways and role of ligninases for the degradation of lignin substructure models by white rot fungi. Wood Res. 73:58–81.

225 Higuchi, T., and Y. Nakamura. 1978. Ester linkage of p–coumaric acid in Bamboo lignin. Cell. Chem. Technol. 12:199–208.

226 Hinton, D.M., and Bacon, C.W. 1985. The distribution and ultrastructure of the endophyte of toxic tall fescue. Can. J. Bot. 63:36–42.

227 Hobot, J. A. 1989. Lowicryls and low-temperature embedding for colloidal gold methods.Pages 76–112 in: Colloidal Gold: Principles, Methods, and Applications. Vol. 2. M. A. Hayat, ed. Academic Press, New York.

228 Hoch, H. C., and Staples, R. C. 1987. Structural and chemical changes among the rust fungi during appressorium development. Annu. Rev. Phytopathol. 25:231–247.

229 Hoch, H. C., and Staples, R. C. 1991. Signaling for infection structure formation in fungi. Pages 25–46 in: The fungal Spore and Disease Initiation in Plants and Animals. G. T. Cole and H. C. Hoch, eds. Plenum Press, New York.

230 Holloway, P. J. 1982. The chemical constitution of plant cutins. Pages 45–85 in: The Plant Cuticle. D. F. Cutler, K. L. Alvin and C. E. Price, eds. Academic Press, London.

231 Honegger, R. 1985. Scanning electron microscopy of the fungus-plant cell interface: a simple preparative technique. Trans. Br. mycol. Soc. 84:530–533.

232 Horisberger, M. 1979. Evaluation of colloidal gold as a cytochemical marker for transmission and scanning electron microscopy. Biol. Cell 36:253–258.

233 Howard, R. J., Ferrari, M. A., Roach, D. H., and Money, N. P. 1991. Penetration of hard substrates by a fungus employing enormous turgor pressures. Proc. Natl. Acad. Sci. USA 88:11281–11284.

234 Humphreys, A. M., and Gooday, G. W. 1984. Phospholipid requirement of microsomal chitinase from *Mucor mucedo*. Curr. Microbiol. 11:187–190.

235 Humphreys, A. M., and Gooday, G. W. 1984. Properties of chitinase activities from *Mucor mucedo*: evidence for membrane–bound zymogenic form. J. Gen. Microbiol. 130:1359–1366.

236 Illman, B. L., and Highley, T. L. 1989. Decomposition of wood by brown–rot fungi. Pages 465–484 in: Biodeterioration Research II. C. E. O'Rear and G. C. Llewellyn, eds. Plenum Press, New York.

237 Iten, W., and Matile, P. 1970. Role of chitinase and other lysosomal enzymes of *Coprinus lagopus* in the autolysis of fruiting bodies. J. Gen. Microbiol. 61:301–309.
238 Jach, G., Logemann, S., Wolb, G., Oppenheim, A., Chet, I., Schell, J., and Logemann, J. 1992. Expression of a bacterial chitinase leads to improved resistance of transgenic tobacco plants against fungal infection. Biopractice 1:33–40.
239 Jarstfer, A. G., and Sylvia, D.M. 1992. Inoculum production and inoculation strategies for vesicular–arbuscular fungi. Pages 349–378 in: Soil Microbial Ecology. F. Blaine Metting, ed. M. Dekker, New York.
240 Jellison, J., Chandhoke, V., Goodell, B., and Fekete, F. A. 1991. The isolation and immunolocalization of iron–binding compounds produced by *Gloeophyllum trabeum*. Appl. Microbiol. Biotech. 35:805–809.
241 Jensen, W. A. 1962. Botanical histochemistry. W.W. Freeman and Co., San Francisco.
242 Jeuniaux, C. 1982. La chitine dans le règne animal. Bull. Soc. Zool. France 107:363–386.
243 Jewell, F. F. Sr. 1988. Histopathology of fusiform rust–inoculated progeny from (shortleaf x slash) x shortleaf pine crosses. Phytopathol. 78:396–402.
244 Johnston, H. W., and Sproston, T. Jr. 1965. The inhibition of fungus infection pegs in *Ginkgo biloba*. Phytopathol. 55:225–227.
245 Joseleau, J. P., and Ruel, K. 1989. Enzyme excretion during wood cell wall degradation by *Phanerochaete chrysosporium*. Pages 443–453 in: Plant Cell Wall Polymers and Biodegradation. N. G. Lewis and M. G. Paice, eds. American Society Series 399, Washington, D.C.
246 Joseleau, J. P., and Ruel, K. 1992. Ultrastructural examinations of lignin and polysaccharide degradation in wood by white-rot fungi. Pages 195–202 in: Biotechnology in Pulp and Paper Industry. M. Kuwahara and M. Shimada, eds. Uni Publishers Co, Ltd, Tokyo.
247 Joseleau, J. P., Comtat, J., and Ruel, K. 1992. Chemical structure of xylans and their interaction in the plant cell walls. Pages 1–16 in: Progress in Biotechnology 7. Xylans and Xylanases. Visser, J. et al. (eds). Elsevier, Amsterdam.
248 Joseleau, J. P., and Kesraoui, R. 1986. Glycosidic bonds between lignin and carbohydrates. Holzforschung 40:163–168.
249 Kämper, J. T., Kämper, U., Rogers, L. M., and Kolattukudy, P.E. 1994. Identification of regulatory elements in cutinase promoter from *Fusarium solani* f. sp. *pisi* (*Nectria haematococca*), J. Biol. Chem., in press.
250 Kandasamy, M. K., Parthasarathy, M. V., and Nasrallah, M. E. 1991. High pressure freezing and freeze substitution improve immunolabelling of S-locus specific glycoproteins in the stigma papillae of *Brassica*. Protoplasma 162:187–191.
251 Katz, G. 1965. The location and significance of O–acctyl groups in a glucomannan from Parana pinc. Dissertation Abstr. Appleton 25:1571–1572.
252 Kcon, J. P. R., Byrde, R. J., Byrde, W., and Cooper, R. M. 1987. Some aspects of fungal enzymes that degrade plant cell walls. Pages 133–157 in: Fungal Infection of Plants. G. F. Pegg and P. G. Ayres, eds. Cambridgc University Press, Cambridge.
253 Kersten, P. J., and Kirk, T. K. 1987. Involvement of a new enzyme, glyoxal oxidase, in extracellular H_2O_2 production by *Phanerochaete chrysosporium*. J. Bacteriol. 169:2195–2201.
254 Kessler, K. J., Jr. 1966. Xylem sap as a growth medium for four tree wilt

fungi. Phytopathol. 56:1165–1169.
255 Kirk, T. K., and Farrell, R. L. 1987. Enzymatic combustion: The microbial degradation of lignin. Annu. Rev. Microbiol. 41:465–505.
256 Kirk, T. K., and Hammell, K. E. 1992. What is the primary agent of lignin degradation in white–rot fungi. Pages 535–540 in: Biotechnology in Pulp and Paper Industry. M. Kuwahara and M. Shimada, eds. Uni Publishers Co, Ltd, Tokyo.
257 Kirk, T. K., Ibach, R., Mozuch, M. D., Conner, A. H., and Highley, T. L. 1991. Characteristics of cotton cellulose depolymerized by a brown–rot fungus, by acid, or by chemical oxidants. Holzforschung 45:239–244.
258 Knox, J. P. 1990. Emerging patterns of organization at the plant cell surface. J. Cell Sci. 96:557–561.
259 Knox, J. P., Linstead, P. J., King, J., Cooper, C., and Roberts, K. 1990. Pectin esterification is spatially regulated both within cell walls and between developing tissues of root apices. Planta 181:512–521.
260 Kolattukudy, P. E. 1980. Biopolyester membranes of plants: cutin and suberin. Science 208:990–1000.
261 Kolattukudy, P. E. 1980. Cutin, suberin and waxes. Pages 571–645 in: Comprehensive Biochemistry of Plants, Vol. 4, P. K. Stumpf, ed. Academic Press, London.
262 Kolattukudy, P. E. 1981. Structure, biosynthesis and biodegradation of cutin and suberin. Annu. Rev. Plant Physiol. 32: 539–567.
263 Kolattukudy, P. E. 1984. Fungal penetration of defensive barriers of plants. Pages 302–343 in: Structure, Function and Biosynthesis of Plant Cell Walls. W. M. Dugger and S. Bartnicki–Garcia, eds. American Society of Plant Physiologists, Waverly Press, Baltimore, MD.
264 Kolattukudy, P. E. 1984. Cutinases from fungi and pollen. Pages 472–504 in: Lipases. B. Borgstrom and H.L. Brockman, eds. Elsevier, Amsterdam.
265 Kolattukudy, P. E. 1985. Enzymatic penetration of the plant cuticle by fungal pathogens. Annu. Rev. Phytopathol. 23:223–250.
266 Kolattukudy, P. E. 1987. Lipid derived polymers and waxes and their role in plant–microbe interaction. Pages 291–314 in: The Biochemistry of Plants, Vol. 9, P. K. Stumpf, ed. Academic Press, London.
267 Kolattukudy, P. E., and Espelie, K. E. 1985. Biosynthesis of cutin, suberin, and associated waxes. Pages 161–207 in: Biosynthesis and biodegradation of wood components. T. Higuchi, ed. Academic Press, Orlando.
268 Kolattukudy, P. E., and Espelie, K. E. 1989. Chemistry, biochemistry and function of suberin and associated waxes. Pages 304–367 in: Natural Products of Woody Plants, Chemicals Extraneous to the Lignocellulosic Cell Wall. J. Rowe, ed. Springer–Verlag.
269 Kolattukudy, P. E., Ettinger, W. F., and Sebastian, J. 1987. Cuticular lipids in plant–microbe interactions. Pages 473–480 in: The Metabolism, Structure and Function of Plant Lipids. P. K. Stumpf, B. D. Mudd and W. D. Ness, eds. Plenum Publishing Corporation, New York and London.
270 Kolattukudy, P. E., Mohan, R., Bajar, A. M., and Sherf, B. A. 1992. Plant peroxidase gene expression and function. Biochem. Soc. Trans. 20:333–337.
271 Kolattukudy, P. E., Podila, G. K., Roberts, E., and Dickman, M. B. 1989. Gene expression resulting from the early signals in plant–fungus interaction. Pages 87–102 in: Molecular Biology of Plant–Pathogen Interactions. UCLA Symposia on Molecular and Cellular Biology, New

series, Vol 101. B. Staskawicz, P. Ahlquist and O. Yoder, eds. Alan R. Liss, Inc. New York.
272 Kolattukudy, P. E., Podila, G. K., Sherf, B. A., Bajar, A., and Mohan, R. 1991. Mutual triggering of gene expression in plant–fungus interactions. Pages 242–249 in: Advances in Molecular Genetics of Plant–Microbe Interactions, Vol. 1. H. Hennecke and D. P. S. Verma, eds. Kluwer Academic Publishers, The Netherlands.
273 Kolattukudy, P. E., and Walton, T. J. 1973. The biochemistry of plant cuticular lipids. Prog. Chem. Fats and Other Lipids 13:121–175.
274 Köller, W. 1991. Plant cuticles: the first barrier to be ovecome by plant pathogens. Pages 219–246 in: The Fungal Spore and Disease Initiation in Plants and Animals. G. T. Cole and H. C. Hoch, eds. Plenum Press, New York.
275 Köller, W., and Kolattukudy, P. E. 1982. Role of cutinase and cell wall degrading enzymes in infection of *Pisum sativum* by *Fusarium solani* f.sp. *pisi*. Physiol. Plant Pathol. 20:47–60.
276 Köller, W., Parker, D. M., and Becker, C. M. 1991. Role of cutinase in the penetration of apple leaves by *Venturia inaequalis*. Phytopathol. 81: 1375– 1379.
277 Koshijima, T., Taniguchi, T., and Tanaka, R. 1972. Lignin carbohydrate complex. The influences of milling of wood upon the Björkman LCC. Holzforschung 26:211–217.
278 Kottke, I., and Oberwinkler, F. 1986. Mycorrhiza of forest trees – structure and function. Trees 1:1–24.
279 Kraft, J. M., Burke, D. W., and Haglund, W. A. 1981. *Fusarium* diseases of beans, peas and lentils. Pages 142–156 in: *Fusarium*: Diseases, Biology and Taxonomy. P. E. Nelson, T. A. Toussoun and R. J. Cook, eds. The Pennsylvania State University Press, University Park 198.
280 Kremer, S. M., and Wood, P. M. 1992. Evidence that cellobiose oxidase from *Phanerochaete chrysosporium* is primarily an Fe(III) reductase. Eur. J. Biochem. 205:133–138.
281 Kubo, Y., and Furusawa, I. 1991. Melanin biosynthesis. Prerequisite for successful invasion of the plant host by appressoria of *Colletotrichum* and *Pyricularia*. Pages 205–218 in: The Fungal Spore and Disease Initiation in Plants and Animals. G. T. Cole and H. C. Hoch, eds. Plenum Press, New York.
282 Kunz, C., Sellan, O., and Bertheau, Y. 1992. Purification and characterization of a chitinase from the hyperparasitic fungus *Aphanocladium album*. Physiol. Mol. Plant. Pathol. 40:117–131.
283 Kuranda, M. J., and Robbins, P.W. 1991. Chitinase is required for cell separation during the growth of *Saccharomyces cerevisiae*. J. Biol. Chem. 266:19758–19767.
284 Lambais, M. R., and Medhy, M. C. 1993.Suppression of endochitinase, β 1–3–endoglucanase, and chalcone isomerase expression in bean vesicular–arbuscular mycorrhizal roots under different soil phosphate conditions. Mol. Plant-Microbe Interact. 6:75–83.
285 Larsen, M. J., and Green III, F. 1992. Mycofibrillar cell wall extensions in the hyphal sheath of *Postia placenta*. Can. J. Microbiol. 38:905–911.
286 Lawton, M. A., and Lamb, C. J. 1987. Transcriptional activation of plant defense genes by fungal elicitor, wounding and infection. Mol. Cell. Biol. 7:335-341.
287 Leary, G. J., Santell, D. A., and Wong, H. 1983. The formation of model lignin–carbohydrate compounds in aqueous solution. Holzforschung

37:11–16.
288 Leatham, G. F. 1986. The ligninolytic activities of *Lentinus edodes* and *Phanerochaete chrysosporium*. Appl. Microbiol. Biotechnol. 24:51–58.
289 Lei, J., and Dexheimer, J. 1988. Ultrastructural localization of ATPase activity in the *Pinus sylvestris–Laccaria laccata* ectomycorrhizal association. New Phytol. 108:329–334.
290 Leightley, L.E, and Eaton, R.A. 1980. Micromorphology of wood decay by marine microorganisms. Pages 83–88 in: Biodeterioration Symposium Berlin, 1978. Edited by T.A. Oxley, D. Allsopp, and G. Becker.
291 Liese, W., and Schmid, R. 1963. Fibrillarstrukturen an den Hyphen holzzerstörender Pilze. Naturwissenschaften 50:102–103.
292 Liese, W., and Schmid, R. 1964. Ueber das Wachstum von Bläuepilzen durch verholzte Zellwände. Phytopathol. Z. 51:385–393.
293 Lin, S. Y., and Dence, C. W. 1992. Methods in lignin chemistry. Springer–Verlag. Berlin–Heidelberg.
294 Linthorst, H. J. M., Van Loon, L. C., Memelink, J., and Bol, J. F. 1990. Characterization of cDNA clones for a virus-inducible, glycine-rich protein from petunia. Plant Mol. Biol. 15:521-523.
295 Linthorst, H. J. M., Van Loon, L. C., Van Rossum, C. M. A., Mayer, A., Bol, J. F., Van Roekel, J. S. C., Meulennoff, E. J. S., and Cornelissen, B. J. C. 1990. Analysis of acidic and basic chitinases from tobacco and petunia and their constitutive expression in transgenic tobacco. Mol. Plant–Microbe Interact. 3:252–258.
296 Littlefield, L. J., and Bracker, C. E. 1972. Ultrastructural specialization of the host pathogen interface in rust-infected flax. Protoplasma 74:271–305.
297 Lowerts, G. A., and Kellison, R. C. 1981. Genetically controlled resistance to discoloration and decay in wounded trees of yellow-poplar. Silvae Genet. 30:98–101.
298 Lulai, E. C., and Morgan, W. C. 1992. Histochemical probing of potato periderm with neutral red: a sensitive cytofluorochrome for hydrophobic domain of suberin. Biotech. Histochem. 67:185–195.
299 Lund, P., Lee, R. Y., and Dunsmuir, P. 1989. Bacterial chitinase is modified and secreted in transgenic tobacco. Plant Physiol. 91:131–135.
300 Luttrell, E. S. 1974. Parasitism of fungi in vascular plants. Mycologia 66:1–15.
301 Mahadevan, P. R., and Mahadkar, U. R. 1970. Role of enzymes in growth and morphology of *Neurospora crassa*: Cell wall bound enzymes and their possible role in branching. J. Bacteriol. 101:941–947.
302 Mahan, M. J., Slauch, J. M., and Mekalanos, J. J. 1993. Selection of bacterial virulence genes that are specifically induced in host tissues. Science 259:686–688.
303 Manners, J. M., and Gay, J. L. 1983. The host–parasite interface and nutrient transfer in biotrophic parasitism. Pages 163–195 in: Biochemical Plant Pathology. J. A. Callow, ed. Wiley, New York.
304 Manocha, M. S. 1987. Cellular and molecular aspects of fungal host–mycoparasite interaction. J. Pl. Dis. Prot. 94:431–444.
305 Manocha, M. S., and Balasubramanian, R. 1988. *In vitro* regulation of chitinase and chitin synthetase activity of two mucoraceous hosts of a mycoparasite. Can. J. Microbiol. 34:1116–1121.
306 Manocha, M. S., and Chen, Y. 1990. Specificity of attachment of fungal parasites to their hosts. Can. J. Microbiol. 36:69–76.
307 Manocha, M.S., and Golesorkhi, R. 1981. Host–parasite relations in a mycoparasite. VII. Light and scanning electron microscopy of interactions

of *Piptocephalis virginiana* with host and non–host species. Mycologia 73:976–987.
308 Mansfield, J. W., and Hutson, R. A. 1980. Microscopical studies on fungal development and host responses in broad bean and tulip leaves inoculated with five species of *Botrytis*. Physiol. Plant Pathol. 17:131–144.
309 Mansfield, J. W., and Richardson, A. 1981. The ultrastructure of interactions between *Botrytis* species and broad bean leaves. Physiol. Plant Pathol. 19:41–48.
310 Marinez, C., De Geus, P., Lauwereys, M., Matthyssens, G., and Cambillau, C. 1992. *Fusarium solani* cutinase is a lipolytic enzyme with a catalytic serine accessible to solvent. Nature 356:615–618.
311 Mauch, F., and Staehelin, L. A. 1989. Functional implications of the subcellular localization of ethylene-induced chitinase and β-1,3-glucanase in bean leaves. Plant Cell 1:447–457.
312 Mazau, D., Rumeau, D., and Esquerré-Tugayé, M. T. 1986. Biochemical study of hydroxypoline-rich glycoproteins in plant-pathogen interactions. Pages 245-251; in: Biology and Molecular Biology of Plant-Pathogen Interactions. J. A. Bailey, ed. Springer-Verlag, Berlin.
313 McBride, R. P., and Hayes, A. J. 1979. Interactions of the leaf pathogen *Meria laricis* with nutrients and microorganisms of the larch phylloplane. Nova Hedwigia 31:507–517.
314 McCutcheon, T. L., Carroll, G. C., and Schwab, S. 1993. Genotypic diversity in populations of a fungal endophyte from Douglas-fir. Mycologia 85:180–186.
315 McFadden, G. I. 1991. *In situ* hybridization techniques: molecular cytology goes ultrastructural. Pages 219–256 in: Electron Microscopy of Plant Cells. L. Hall and C. Hawes, eds. Academic Press, New York.
316 McNabb, H. S. Jr., Heybroek, H. M., and MacDonald, W. L. 1970. Anatomical factors in resistance to Dutch elm disease. Neth. J. Plant Pathol. 76:196–204.
317 McRae, C. F., and Stevens, G. R. 1990. Role of conidial matrix of *Colletotrichum orbiculare* in pathogenesis of *Xanthium spinosum*. Mycol. Res. 94:890–896.
318 Mehdy, M. C., and Lamb, C. J. 1987. Chalcone isomerase cDNA cloning and mRNA induction by fungal elicitor, wounding and infection. EMBO J. 6:1527–1533.
319 Meikle, P. J., Bonig, I., Hoogenraad, N. J., Clarke, A. E., and Stone, B. A. 1991. The location of (1-3)-β-glucans in the walls of pollen tubes of *Nicotiana alata* using a (1-3)-β-glucan specific monoclonal antibody. Planta 185:1–8.
320 Mellon, J. E., and Helgeson, J. P. 1982. Interaction of a hydroxyproline-rich protein from tobacco callus with potential pathogens. Plant Physiol. 70:401-405.
321 Mendgen, K., and Deising, H. 1993. Infection structures of fungal plant pathogens - a cytological and physiological evaluation. New Phytol. 124:193–213.
322 Mendgen, K., Schneider, A., Sterk, M., and Finck, W. 1988. The differentiation of infection structures as a result of recognition events between some biotrophic parasites and their hosts. J. Phytopathol. 123:259–272.
323 Mendgen, K., Welter, K., Scheffold, F., and Knauf–Beiter, G. 1990. High pressure freezing of rust infected plant leaves. Pages 31–42 in: Electron Microscopy of Plant Pathogens. K. Mendgen and D. E. Leseman, eds. Springer Verlag, Berlin.

324 Merrill, W. 1992. Mechanisms of resistance to fungi in woody plants: a historical perspective. Pages 1–12 in: Defense mechanisms of woody plants against fungi. R. A. Blanchette and A. R. Biggs, eds. Springer–Verlag, Berlin.
325 Milewski, S., O'Donnell, R. W., and Gooday, G. W. 1992. Chemical modification studies of the active centre of *Candida albicans* chitinase and its inhibition by allosamidin. J. Gen. Microbiol. 138:2545–2550.
326 Millar, D. J., Slabas, A. R., Sidebottom, C., Smith, C. G., Allen, A. K., and Bolwell, G. P. 1992. A major stress inducible Mr-42000 wall glycoprotein of French bean (*Phaseolus vulgaris* L.). Planta 187:176–184.
327 Mims, C. W. 1991. Using electron microscopy to study plant pathogenic fungi. Mycologia 83:1–19.
328 Minter, D. W., Staley, J. M., and Millar, C. S. 1978. Four species of *Lophodermium* on *Pinus sylvestris*. Trans. Br. mycol. Soc. 71:295–301.
329 Mohan, R., Bajar, A. M., and Kolattukudy, P. E. 1993. Induction of a tomato anionic peroxidase gene (tap1) by wounding in transgenic tobacco and activation of tap1/GUS and tap2/GUS chimeric gene fusions in transgenic tobacco by wounding and pathogen attack. Plant. Mol. Biol. 21:341–354.
330 Mohan, R., and Kolattukudy, P. E. 1990. Differential activation of expression of a suberization–associated anionic peroxidase gene in near–isogenic resistant and susceptible tomato lines by elicitors of *Verticillium albo–atrum*. Plant Physiol. 921:276–280.
331 Mohan, R., Vijayan, P., and Kolattukudy, P. E. 1993. Developmental and tissue specific expression of a tomato anionic peroxidase (tap1) gene by a minimal promoter, with wound and pathogen induction by an additional 5'–flanking region. Plant Molec. Biol. 22:475–490.
332 Mondal, A. H., and Parbery, D. G. 1992. The spore matrix and germination in *Colletotrichum musae*. Mycol. Res. 96:592–596.
333 Moore, P. J., Darvill, A. G., Albersheim, P., and Staehelin, L. A. 1986. Immunogold localization of xyloglucan and rhamnogalacturonan I in the cell walls of suspension-cultured sycamore cells. Plant Physiol. 82:787–794.
334 Mora, F., Ruel, K., Comtat J., and Joseleau, J. P. 1986. Aspect of native and redeposited xylans at the surface of cellulose microfibrils. Holzforschung 40:85–90.
335 Morelet, M. 1989. L'anthracnose des chênes et du hêtre en France. Rev. Forest. Franç. 41:488–496.
336 Murmanis, L., Highley, T. L., and Palmer, J. G. 1984. An electron microscopy study of western hemlock degradation by the white-rot fungus *Ganoderma applanatum*. Holzforschung 38:11–18.
337 Nakano, J., and Meshitsuka, G. 1992. The detection of lignin. Pages 23–32 in: Methods in Lignin Chemistry. S. Y. Lin and C. W. Dence, eds. Springer–Verlag, Berlin–Heidelburg.
338 Neuhaus, J. M., Ahi–Goy, P., Hinz, U., Flores, S., and Meins, F., Jr. 1991. High level expression of a tobacco chitinase gene in *Nicotiana silvestris*: susceptibility of transgenic plants to *Cercospora nicotianae*. Plant Mol. Biol. 16:141–151.
339 Newman, G. R., and Hobot, J. A. 1989. Role of tissue processing in colloidal methods. Pages 33–43 in: Colloidal Gold. Principles, Methods, and Applications.Vol. 2. M.A. Hayat, ed. Academic Press, New York.
340 Nicholson, R. L., and Epstein, L. 1991. Adhesion of fungi to the plant surface. Prerequiste for pathogenesis. Pages 3–23 in: The Fungal Spore

and Disease Initiation in Plants and Animals. G. T. Cole and H. C. Hoch, eds. Plenum Press, New York.

341 Nicholson, R. L., Hipskind, J., and Hanau, M. 1989. Protection against phenol toxicity by the spore mucilage of *Colletotrichum graminicola*, an aid to secondary spread. Physiol. Mol. Plant Pathol. 35:243–252.

342 Nicole, M., and Benhamou, N. 1991. Cytochemical aspects of cellulose breakdown during the infection process of rubber tree roots by *Rigidoporus lignosus*. Phytopathol. 81:1412–1420.

343 Nicole, M., and Benhamou, N. 1991. Ultrastructural localization of chitin in cell walls of *Rigidoporus lignosus*, the white-rot root fungus of rubber trees. Physiol. Mol. Plant Pathol. 39:415–432.

344 Nicole, M., Chamberland, H., Geiger, J. P., Lecours, N., Valéro, J., Rio, B., and Ouellette, G. B. 1992. Immunocytochemical localization of laccase in wood decayed by *Rigidoporus lignosus*. Appl. Environ. Microbiol. 58:1727–1739.

345 Nicole, M., Chamberland, H., Nandris, D., and Ouellette G. B. 1992. Cellulose is degraded during phloem necrosis of *Hevea brasiliensis*. Eur. J. For. Path. 22:266–277.

346 Nicole, M., Chamberland, H., Rioux, D., Geiger, J. P., Rio, B., Lecours, N., and Ouellette, G. B. 1993. A cytochemical study of extracellular sheaths associated with *Rigidoporus lignosus* during wood decay. Appl. Environ. Microbiol. 59:2578–2588.

347 Nicole, M., Geiger, J. P., and Nandris, D. 1987. Ultrastructural aspects of rubber tree root rot diseases. Eur. J. For. Pathol. 17:1–10.

348 Nieduszynski, I. A., and Marchessault, R.H. 1972. Structure of β–D (1-4)–xylan hydrate. Biopolymers 11:1335–1344.

349 Nieves, R. A., Ellis, R. P., Todd, R. A., Johnson, T. J. A., Grohman, K., and Himmel, M. E. 1991. Visualization of *Trichoderma reesei* cellobiohydrolase I and endo–glucanase I on aspen cellulose by using monoclonal antibody–colloid gold conjugates. Appl. Environ. Microbiol. 57:3163–3170.

350 Nilsson, T., Daniel, G., Kirk, T. K., and Obst, J. R. 1989. Chemistry and microscopy of wood decayed by some higher ascomycetes. Holzforschung 43:11–18.

351 Nimz, H. H. 1974. Beech lignin – proposal of a constitutional scheme. Angew. Chem. Int. Ed. 13:313–321.

352 Northcote, D. H., Davey, R., and Lay, J. 1989. Use of antisera to localize callose, xylan and arabinogalactan in the cell–plate, primary and secondary walls of plant cells. Planta 178:353–366.

353 Nylund, J. E. 1987. The ectomycorrhizal infection zone and its relation to acid polysaccharides of cortical cell walls. New Phytol. 106:505–516.

354 O'Connell, R. J., Brown, I. R., Mansfield, J. W., Bailey, J. A., Mazau, D., Rumeau, D., and Esquerré-Tugayé, M. T. 1990. Immunocytochemical localization of hydroxyproline-rich glycoproteins accumulating in melon and bean at sites of resistance to bacteria and fungi. Mol. Plant-Microbe Interact. 3:33-40.

355 Obst, J. R. 1982. Guaiacyl and syringyl lignin composition in hardwood cell components. Holzforschung 36:143–152.

356 Ong, E., Greenwood, J. M., Gilkes, N. R., Kilburn, D. G., Miller, R. C., Jr., and Warren, R. A. J. 1989. The cellulose–binding domains of cellulases: tools for biotechnology. Trends Biotechnol. 7:239–243.

357 Ordentlich, A., Elan, Y., and Chet, I. 1988. The role of chitinase of *Serratia marcescens* in biocontrol of *Sclerotium rolfsii*. Phytopathol.

78:84–88.
358 Ordentlich, A., Migheli, Q., and Chet, I. 1991. Biological control activity of three *Trichoderma* isolates against *Fusarium* wilts of cotton and muskmelon and lack of correlation with their lytic enzymes. J. Phytopathol. 133:177–186.
359 Ouellette, G. B. 1962. Morphological characteristics of *Ceratocystis ulmi* (Buism.) C. Moreau in American elm trees. Can. J. Bot. 40:1463–1466.
360 Ouellette, G. B., and Benhamou, N. 1987. Use of monoclonal antibodies to detect molecules of fungal plant pathogens. Can. J. Plant Pathol. 9:167–176.
361 Ouellette, G. B., and Rioux, D. 1992. Anatomical and physiological aspects of resistance to Dutch elm disease. Pages 257–307 in: Defense Mechanisms of Woody Plants against Fungi. R. A. Blanchette and A. R. Biggs, eds. Springer-Verlag, Berlin.
362 Ouellette, G. B., and Rioux, D. 1993. Alterations of vessel elements and reactions of surrounding tissues in the DED syndrome. Pages 255-292 in: Dutch Elm Disease Research: Cellular and Molecular Approaches. M. B. Sticklen and J. L. Sherald, eds.. Springer-Verlag New York, etc.
363 Palmer, J. G., Murmanis, L., and Highley, T. L. 1983b. Visualization of hyphal sheath in wood decay Hymenomycetes. II. White-rotters. Mycologia 75:1005–1009.
364 Parker, A. K., and Reid, J. 1969. The genus *Rhabdocline* Syd. Can. J. Bot. 47:1533–1545.
365 Paszczynski, A., Crawford, R. L., and Blanchette, R. A. 1988. Delignification of wood chips and pulps by using natural and synthetic porphyrins: Models of fungal decay. Appl. Environ. Microbiol. 54:62–68.
366 Pawley, J. 1989. The handbook of biological confocal microscopy. Papers given at the confocal workshop. The Electron Microscope Society of America. August 1989, San Antonio, Texas. IMR Press, Madison.
367 Pearce, R. B., and Rutherford, J. 1981. A wound–associated suberized barrier to the spread of decay in the sapwood of oak (*Quercus robur* L.). Physiol. Plant Pathol. 19:359–369.
368 Pearce, R. B., and Woodward, S. 1986. Compartmentalization and reaction zone barriers at the margin of decayed sapwood in *Acer saccharinum* L. Physiol. Mol. Plant Pathol. 29:197–216.
369 Peek, R. D., Liese, W., and Parameswaran, N. 1972. Infektion und Abbau des Wurzelholzes von Fichte durch *Fomes annosus*. Eur. J. For. Pathol. 2:237–248.
370 Peek, R. D., Liese, W., and Parameswaran, N. 1972. Infektion und Abbau der Wurzelrinde von Fichte durch *Fomes annosus*. Eur. J. For. Pathol. 2:104–115.
371 Pegg, G. F. 1988. Chitinase from *Verticillium albo–atrum*. Pages 474–479 in: Methods in Enzymol. Vol. 161 part B. W. A. Wood and S. T. Kellogg, eds. Academic Press, New York, London.
372 Pelissier, B., Thibaud, J. B., Grignon, C., and Esquerré–Tugayé, M. T. 1986. Cell surfaces in plant–microorganism interactions. VII. Elicitor preparations from two fungal pathogens depolarize plant membranes. Plant Science 46:103–109.
373 Peretto, R., Bettini, V., and Bonfante, P. 1993. Evidence of two polygalacturonases produced by a mycorrhizal ericoid fungus during its saprophytic growth. FEMS Microbiology Letters 114:85–92.
374 Perie, F. H., and Gold, M. H. 1991. Manganese regulation of manganese peroxide expression and lignin degradation by the white rot fungus

Dichomitus squalens. Appl. Environ. Microbiol. 57:2240–2245.
375 Perotto, S., Brewin, N., and Bonfante, P. 1993. Colonisation of pea roots by arbuscular mycorrhizal fungi and rhizobia: an immunological comparison using monoclonal antibodies as probes for plant cell surface components. Mol. Plant–Microbe Interact. 6:745–754.
376 Peterson, R. L., and Bonfante, P. 1994. Comparative structure of arbuscular mycorrhizas and ectomycorrhizas. Plant and Soil 159, in press.
377 Petrini, O. 1986. Taxonomy of endophytic fungi of aerial plant tissues. Pages 175–187 in: Microbiology of the Phyllosphere. N. J. Fokkema and J. van den Heuvel, eds. Cambridge University Press, Cambridge.
378 Petrini, O. 1987. Endophytic fungi of alpine Ericaceae. The endophytes of *Loiseleuria procumbens*. Pages 71–77 in: Arctic and Alpine Mycology II. G. A. Laursen, J. F. Ammirati and S.A. Redhead, eds. Environmental Science Research Vol. 34, Plenum Press, London.
379 Petrini, O. 1991. Fungal endophytes of tree leaves. Pages 179–197 in: Microbial Ecology of Leaves. J. H. Andrews and S. S. Hirano, eds. Springer Verlag, New York, Berlin, Heidelberg.
380 Pham, T. T. T., Maaroufi, A., and Odier, E. 1990. Inheritance of cellulose– and lignin–degrading ability as well as endoglucanase isozyme pattern in *Dichomitus squalens*. Appl. Microbiol. Biotechnol. 33:99–104.
381 Philipson, M. N. 1989. A symptomless endophyte of ryegrass (*Lolium perennne*) that spores on its host – a light microscope study. N. Zeal. J. Bot. 27:513–519.
382 Philipson, M. N. 1991. Ultrastructure of the *Gliocladium*–like endophyte of perennial ryegrass (*Lolium perenne* L.). I. Vegetative phase and leaf blade sporulation. New Phytol. 117:271–280.
383 Philipson, M. N. 1991. Ultrastructure of the *Gliocladium*–like endophyte of perennial ryegrass (*Lolium perenne* L.). II. Sporulation in the leaf sheath. New Phytol. 117:281–288.
384 Philipson, M. N. 1991. Ultrastructure of a symptomless fungal endophyte of *Festuca arundinacea*. Bot. Gaz. 152:296–303.
385 Philipson, M. N., and Christey, M. C. 1985. An epiphytic/ endophytic fungal associate of *Danthonia spicata* transmitted through the embryo sac. Bot. Gaz. 146:70–81.
386 Philipson, M. N., and Christey, M. C. 1986. The relationship of host and endophyte during flowering, seed formation, and germination of *Lolium perenne*. New Zeal. J. Bot. 24:125–134.
387 Pichè, Y., Peterson, R. L., and Massicotte, H. B. 1988. Host–fungus interactions in ectomycorrhizae. Pages 55–71 in: Cell to Cell Signals in Plant, Animal, and Microbial Symbiosis. S. Scannerini, D. C. Smith, P. Bonfante and V. Gianinazzi, eds. NATO ASI H 17 Springer Verlag, Berlin.
388 Pielken, P., Stahmann, P., and Sahm, H. 1990. Increase in glucan formation by *Botrytis cinerea* and analysis of the adherent glucan. Appl. Microbiol. Biotechnol. 33:1–6.
389 Podila, G. K., Dickman, M. B., and Kolattukudy, P. E. 1988. Transcriptional activation of a cutinase gene in isolated fungal nuclei by plant cutin monomers. Science 242:922–925.
390 Podila, G. K., Dickman, M. B., Rogers, L. M., and Kolattukudy, P. E. 1989. Regulation of expression of fungal genes by plant signals. Pages 217–226 in: Molecular Biology of Filamentous Fungi, H. Nevalainen and M. Penttila, eds. Foundation for Biotechnical and Industrial Fermentation Research 6.

391 Politis, D. J. 1976. Ultrastructure of penetration by *Colletotrichum graminicola* of highly resistant oat leaves. Physiol. Plant Pathol. 8:117–122.
392 Politis, D. J., and Wheeler, H. 1973. Ultrastructural study of the penetration of maize leaves by *Colletotrichum graminicola*. Physiol. Plant Pathol. 3:465–471.
393 Poon, N. H., and Day, A. W. 1975. Fungal fimbriae. I. Structure, origin, and synthesis. Can. J. Microbiol. 21:537–546.
394 Popp, J. L., Kalyanaraman, M., and Kirk, T. K. 1990. Lignin peroxidase oxidation of Mn^{+2} in the presence of veratryl alcohol, malonic or oxalic acid, and oxygen. Biochem. 29:10475–10480.
395 Poutanen, K., Puls, J., and Linko, M. 1986. Hydrolysis of steamed birch wood hemicellulose by enzymes produced by *Trichoderma reesei* and *Aspergillus awamori*. Appl. Microbiol. Biotechnol. 23:487–490.
396 Purdy, R. E., and Kolattukudy, P. E. 1975. Hydrolysis of plant cuticle by pathogens. Properties of cutinase I, cutinase II and a nonspecific esterase isolated from *Fusarium solani pisi*. Biochem. 14:2832–2840.
397 Purdy, R. E., and Kolattukudy, P. E. 1975. Hydrolysis of plant cuticle by pathogens. Purification, amino acid composition, and molecular weight of two isozymes of cutinase and a nonspecific esterase from *Fusarium solani* f. *pisi*. Biochem. 14:2824–2831.
398 Rast, D. M., Horsch, M., Furter, R., and Gooday, G. 1991. A complex chitinolytic system in exponentially growing mycelium of *Mucor rouxii*: properties and function. J. Gen. Microbiol. 137:2797–2810.
399 Read, D. J. 1991. Mycorrhizas in ecosystems. Experientia 47:376–390.
400 Reid, I. D. 1989. Solid state fermentations for biological delignification. Enzyme Microb. Technol. 11:786–803.
401 Renganathan, V., Usha, S. N., and Lindenburg, F. 1990. Cellobiose–oxidizing enzymes from the lignocellulose–degrading basidiomycete *Phanerochaete chrysosporium*: interaction with microcrystalline cellulose. Appl. Microbiol. Biotechnol. 32:609–613.
402 Reyes, F., Calatayud, J., and Martinex, M. J. 1989. Endochitinase from *Aspergillus nidulans* implicated in the autolysis of its cell wall. FEMS Microbiol. Lett. 60:119–124.
403 Rghei, N. A., Castle, A. J., and Manocha, M. S. 1992. Involvement of fimbriae in host mycoparasite interaction. Physiol. Mol. Plant Pathol. 41:139–148.
404 Rice, J. S., Pinkerton, B. W., Stringer, W. C., and Unersander, D. J. 1990. Seed production in tall fescue as affected by a fungal endophyte. Crop Science 30:1303–1305.
405 Ride, J. P. 1983. Cell walls and other structural barriers in defence. Pages 215–236 in: Biochemical Plant Pathology. J. A. Callow, ed. John Wiley and Sons, Chichester.
406 Ride, J. P. 1985. Non–host resistance to fungi. Pages 29–61 in: Mechanisms of Resistance to Plant Diseases. R. S. S. Fraser, ed. Martinus Nijhoff/Dr. W. Junk, Publishers, Boston.
407 Ride, J. P., and Pearce, R. B. 1979. Lignification and papilla formation at sites of attempted penetration of wheat leaves by non–pathogenic fungi. Physiol. Plant Pathol. 15:79–92.
408 Riffle, J. W., and Peterson, G. W. 1986. Thyronectria canker of honey-locust: influence of temperature and wound age on disease development. Phytopathol. 76:313–316.
409 Rioux, D., and Ouellette, G. B. 1989. Light microscope observations of

histological changes induced by *Ophiostoma ulmi* in various nonhost trees and shrubs. Can. J. Bot. 67:2335–2351.

410 Rioux, D., and Ouellette, G. B. 1991. Barrier zone formation in host and nonhost trees inoculated with *Ophiostoma ulmi*. I. Anatomy and histochemistry. Can. J. Bot. 69:2055–2073.

411 Robards, A. W. 1991. Rapid-freezing methods and their application. Pages 257–313 in: Electron Microscopy of Plant Cells. J. L Hall and C. Hawes, eds. Academic Press, New York.

412 Robb, J., Lee, S.–W., Mohan, R., and Kolattukudy, P. E. 1991. Chemical characterization of stress-induced vascular coating in tomato. Plant Physiol. 97:528–536.

413 Roberts, E., and Kolattukudy, P. E. 1989. Molecular cloning, nucleotide sequence and abscisic acid induction of a suberization-associated highly anionic peroxidase. Mol. Gen. Genet. 217:223–232.

414 Roberts, E., Kutchan, T., and Kolattukudy, P. E. 1988. Cloning and sequencing of cDNA for a highly anionic peroxidase from potato and the induction of its mRNA in suberizing potato tuber and tomato fruits. Plant Mol. Biol. 11:15–26.

415 Roberts, K. 1990. Structures at the plant surface. Curr. Opinions Cell Biol. 2:920–928.

416 Rosenberger, R. F. 1979. Endogenous lytic enzymes and wall metabolism. Pages 265–278 in: Fungal Walls and Hyphal Growth. J. H. Burnett and A. P. J. Trinci, eds. Symp. Br. mycol. Soc. Vol. 2, Cambridge University Press, Cambridge.

417 Roth, J. 1978. The lectins: Molecular probes in cell biology and membrane research. Exp. Pathol. (Suppl). 3:1–187.

418 Ruel, K., Ambert, K., and Joseleau, J. P. 1993. Influence of the enzyme equipment of white–rot fungi on the patterns of wood degradation . FEMS Microbiol. Rev. In Press.

419 Ruel, K., Barnoud, F., and Goring, D.A.I. 1978. Lamellation in the S_2 layer of softwood tracheids as demonstrated by scanning transmission electron microscope. Cell. Chem. Technol. 13:429–432.

420 Ruel, K., and Joseleau, J. P. 1984. Use of enzyme-gold complexes for the ultrastructural localization of hemicelluloses in the plant cell wall. Histochem. 81:573–580.

421 Ruel, K., and Joseleau, J. P. 1991. Involvement of an extracellular glucan sheath during degradation of *Populus* wood by *Phanerochaete chrysosporium*. Appl. Environ. Microbiol. 57:374–384.

422 Ruel, K., Odier E., and Joseleau, J. P. 1989. Immunocytochemical observations of fungal enzymes during degradation of wood cells by *Phanerochaete chrysosporium* (strain K-3). Pages 83–97 in: Biotechnology in Pulp and Paper Manufacture. Applications and Fundamental Investigations. T. K. Kirk and H. M. Chang, eds. Butterworth-Heineman, Boston.

423 Russin, J. S., and Shain, L. 1984. Initiation and development of cankers caused by virulent and cytoplasmic hypovirulent isolates of the chestnut blight fungus. Can. J. Bot. 62:2660–2664.

424 Ruttimann, C., Schwember, E., Salas, L., Cullen, D., and Vicuna, R. 1992. Ligninolytic enzymes of the white rot basidiomycetes *Phlebia brevispora* and *Ceriporiopsis subvermispora*. Biotechnol. Appl. Biochem. 16:64–76.

425 Ryder, T. B., Cramer, C. L., Bell, J. N., Robbins, M. P., Dixon, R. A., and Lamb, C. J. 1984. Elicitor rapidly induces chalcone synthase mRNA

in *Phaseolus vulgaris* cells at the onset of phytoalexin defense response. Proc. Natl. Acad.Sci. USA 81:5724-5728.

426 Ryerson, D. E., and Heath, M. C. 1992. Fungal elicitation of wall modifications in leaves of *Phaseolus vulgaris* L. cv. Pinto II. Effects of fungal wall components. Physiol. Mol. Plant Pathol. 40:283–298.

427 Ryser, U., and Keller, B. 1992. Ultrastructural localization of a bean glycine-rich protein in unlignified primary walls of protoxylem cells. Plant Cell 4:773–783.

428 Sahai, A. S., Balasubramanian, R., and Manocha, M. S. 1993. Immunofluorescence study of zygomycetous fungi with two chitin–binding probes. Exp. Mycol. 17:55–69.

429 Saka, S., and Goring, D. A. I. 1988. The distribution of lignin in white birch wood as determined by bromination with TEM–EDXA. Holzforschung 42:149–153.

430 Saka, S., and Thomas, R. J. 1982. Evaluation of the quantitative assay of lignin distribution by SEM–EDXA technique. Wood Sci. Technol. 16:1–18.

431 Samejima, M., Phillips, R. S., and Eriksson, K. E. 1992. Cellobiose oxidase from *Phanerochaete chrysosporium* stopped–flow spectrophotometric analysis of pH–dependent reduction. FEBS 306:165–168.

432 Santamour, F. S. Jr. 1979. Inheritance of wound compartmentalization in soft maples. J. Arboric. 5:220–225.

433 Sarafis, V. 1990. Biological perspectives of confocal microscopy. Pages 325–333 in: Confocal Microscopy. T. Wilson, ed. Academic Press, New York.

434 Sarkanen, K. V., and Hergert, H. L. 1971. Classification and distribution. Pages 43–94 in: Lignins: Occurrence, Formation, Structure and Reactions. K. V. Sarkanen and C. H. Ludwig, eds. Wiley–Interscience, New York.

435 Sauer, N., Corbin, D. R., Keller, B., and Lamb, C. J. 1990. Cloning and characterization of a wound-specific hydroxyproline-rich glycoprotein in *Phaseolus vulgaris*. Plant Cell Environ. 13:257- 266.

436 Scala, C., Cenacchi, G., Ferrari, C., Pasquinelli, G., Preda, P., and Manara, G. C. 1992. A new acrylic resin formulation: A useful tool for histological, ultrastructural and immunocytochemical investigations. J. Histochem. Cytochem. 40:1799–1804.

437 Schmelzer, E., Kruger-Lebus, S., and Hahlbrock, K. 1989. Temporal and spatial patterns of gene expression around sites of attempted fungal infection in parsley leaves. The Plant Cell 1:993–1001.

438 Seifers, D., and Ammon, V. 1980. Mode of penetration of sycamore leaves by *Gloeosporium platani*. Phytopathol. 70:1050–1055.

439 Setliff, E. C., and Hazenberg, M. 1989. Nuclear behavior in the basidiomycete *Rigidoporus vinelus*. Mycol. Soc. America Newsletter 40:47.

440 Shapira, R., Ordentlich, A., Chet, I., and Oppenheim, A. B. 1989. Control of plant diseases by chitinase expressed from cloned DNA in *Escherichia coli*. Phytopathol. 79:1246–1249.

441 Sharp, J. K., Valent, B., and Albersheim, P. 1984. Purification and partial characterization of a β-glucan fragment that elicits phytoalexin accumulation in soybean. J. Biol. Chem. 259:11312–11320.

442 Shaykh, M., Soliday, C., and Kolattukudy, P. E. 1977. Proof for the production of cutinase by *Fusarium solani* f. *pisi* during penetration into its host, *Pisum sativum*. Plant Physiol. 60:170–172.

443 Sheng, J., D'Ovidio, R., and Mehdy, M. C. 1991. Negative and positive regulation of a novel proline-rich protein mRNA by fungal elicitor and

wounding. Plant J. 1:345-354.
444 Sheng, J., Jeong, J., and Mehdy, M. C. 1993. Developmental control and phytochrome-mediated induction of mRNAs encoding a proline-rich protein, hydroxyproline-rich glycoproteins and glycine-rich proteins in *Phaseolus vulgaris* L. Proc. Natl. Acad. Sci. USA 90:828-832.
445 Sherwood, R. T., and Vance, C. P. 1976. Histochemistry of papillae formed in reed canarygrass leaves in response to noninfecting pathogenic fungi. Phytopathol. 66:503–510.
446 Sherwood, R. T., and Vance, C. P. 1980. Resistance to fungal penetration in Gramineae. Phytopathol. 70:273–279.
447 Sherwood–Pike, M., Stone, J. K., and Carroll, G. C. 1986. *Rhabdocline parkeri*, a ubiquitous foliar endophyte of Douglas–fir. Can. J. Bot. 64:1849–1855.
448 Shigo, A. L. 1984. Compartmentalization: a conceptual framework for understanding how trees grow and defend themselves. Annu. Rev. Phytopathol. 22:189–214.
449 Shigo, A. L., and Marx, H. G. 1977. Compartmentalization of decay in trees. Agric. Inf. Bull. USDA For. Serv. No. 405.
450 Shimada, M., Akamatsu, Y., Ma, D. B., and Takahashi, M. 1992. Pages 273–278 in: Biotechnology in Pulp and Paper Industry. M. Kuwahara and M. Shimada (eds). Uni Publishers Co, Ltd, Tokyo.
451 Showalter, A. M. 1993. Structure and function of plant cell wall proteins. Plant Cell 5:9-23.
452 Showalter, A. M., Bell, J. L., Cramer, C. L., Bailey, J. A., Varner, J. E., and Lamb, C. J. 1985. Accumulation of hydroxyproline-rich glycoprotein mRNAs in response to fungal elicitor and infection. Proc. Natl. Acad. Sci. USA 82:6551-6555.
453 Sieber, T. N., and Hugentobler, C. 1987. Endophytische Pilze in Blättern und Ästen gesunder und geschädigter Buchen *(Fagus sylvatica L.)*. Eur. J. For. Pathol. 17:411–425.
454 Sieber-Canavesi, F., and Sieber, T. N. 1988. Endophytische Pilze in Tanne (*Abies alba* Mill.) – Vergleich zweier Standorte im Schweizer Mittelland (Naturwald – Aufforstung). Sydowia 40:250–273.
455 Siegel, M. R., Jarlfors, U., Latch, G. C. M., and Johnson, M. C. 1987. Ultrastructure of *Acremonium coenophialum, Acremonium lolii,* and *Epichloë typhina* endophytes in host and nonhost *Festuca* and *Lolium* species of grasses. Can. J. Bot. 65:2357–2367.
456 Siegel, M. R., and Schardl, C. L. 1991. Fungal endophytes of grasses: detrimental and beneficial associations. Pages 199–221 in: Microbial Ecology of Leaves. J. H. Andrews and S. S. Hirano, eds. Springer Verlag, New York, Berlin, Heidelberg.
457 Sinclair, W. A., Zahand, J. P., and Melching, J. B. 1975. Anatomical marker for resistance of *Ulmus americana* to *Ceratocystis ulmi*. Phytopathol. 65:349–352.
458 Sivan, A., and Chet, I. 1989. Degradation of fungal cell walls by lytic enzymes of *Trichoderma harzianum*. J. Gen. Microbiol. 135:675–682.
459 Sjöström, E. 1993. Wood chemistry. Fundamentals and Application 2nd Ed. Acad. Press, N.Y.
460 Smereka, K. J., Machardy, W. E., and Kausch, P. A. 1987. Cellular differentiation in *Venturia inaequalis* ascospores during germination and penetration of apple leaves. Can. J. Bot. 65:2549–2561.
461 Smith, J., and Grula, E. A. 1983. Chitinase is an inducible enzyme in *Beauveria bassiana*. J. Inv. Pathol. 42:319–326.

462 Smith, S. E., and Smith, F. A. 1990. Structure and function of the biotrophic symbioses as they relate to nutrient transport. New Phytol. 114:1–38.

463 Soliday, C. L., Dickman, M. B., and Kolattukudy, P. E. 1989. Structure of the cutinase gene and detection of promoter activity in the 5'–flanking region by fungal transformation. J. Bacteriol. 171:1942–1951.

464 Sprey, B. 1986. Localisation of β–glucosidase in *Trichoderma reesei* cell walls with immunoelectron microscopy. FEMS Microbiol. Letters 36:287–292.

465 Srebotnik, E., Messner, K., and Foisner, R. 1988. Penetrability of white rot–degraded pine wood by the lignin peroxidase of *Phanerochaete chrysosporium*. Appl. Environ. Microbiol. 54:2608–2614.

466 Srebotnik, E., Messner, K., Foisner, R., and Pettersson, B. 1988. Ultrastructural localization of ligninase of *Phanerochaete chrysosporium* by immunogold labeling. Curr. Microbiol. 16:221–227.

467 St. Leger, R., Cooper, R. M., and Charnley, A. K. 1986. Cuticle degrading enzymes of entomopathogenic fungi: regulation of production of chitinolytic enzymes. J. Gen. Microbiol. 132:1509–1517.

468 St. Leger, R., Cooper, R. M., and Charnley, A.K. 1991. Characterization of chitinase and chitobiase produced by the entomopathogenic fungus *Metarhizium anisopliae*. J. Inv. Pathol. 58:415–426.

469 Stafstrom, J. P., and Staehelin, L. A. 1988. Antibody localization of extensin in cell walls of carrot storage roots. Planta 174:321–332.

470 Stahl, D. J., and Schäfer, W. 1992. Cutinase is not required for fungal pathogenicity on pea. Plant Cell 4: 621–629.

471 Stahmann, K. P., Pielken, P., Schimz, K. L., and Sahm, H. 1992. Degradation of extracellular β-(1,3) (1,6)-D-glucan by *Botrytis cinerea*. Appl. Environ. Microbiol. 10:3347–3354.

472 Stark, R. E., Zlotnik–Mazor, T., Ferrantello, L. M., and Garbow, J. R. 1989. Molecular structure and dynamics of intact plant polyesters. ACS Symposium Series 399:214–229.

473 Stone, J. K. 1986. Foliar endophytes of Douglas fir: cytology and physiology of the host–endophyte relationship. Ph.D. Thesis, University of Oregon, Eugene, Oregon, USA.

474 Stone, J. K. 1987. Initiation and development of latent infections by *Rhabdocline parkeri* on Douglas-fir. Can. J. Bot. 65:2614–2621.

475 Stone, J. K. 1988. Fine structure of latent infections by *Rhabdocline parkeri* on Douglas-fir, with observations on uninfected epidermal cells. Can. J. Bot. 66:45–54.

476 Stone, J. K., Gernandt, D., and Carroll, G. C. 1993. Foliar endophytes of *Taxus brevifolia* Nutt. in western Oregon. International Yew Resources Conference Proceedings. University of California Press. In press.

477 Studer, D., Michel, M., and Muller, M. 1989. High pressure freezing comes of age. Scanning Microsc. Supp. 3:253–269.

478 Stumpf, M. A., and Gay, J. L. 1989. The haustorial interface in a resistant interaction of *Erysiphe pisi* with *Pisum sativum*. Physiol. Mol. Plant Pathol. 35:519–533.

479 Stumpf, M. A., and Heath, M. C. 1985. Cytological studies of the interactions between the cowpea rust fungus and silicon–depleted French bean plants. Physiol. Plant Pathol. 27:369–385.

480 Sundheim, L. 1987. Cloning and conjugational transfer of chitinase encoding genes. J. Agric. Sci. Fin. 59:209–215.

481 Sundheim, L., Poplawsky, A. R., and Ellingboe, A. 1988. Molecular

cloning of two chitinase genes from *Serratia marcescens* and their expression in *Pseudomonas* sp. Physiol. Mol. Plant. Pathol. 33:483–492.
482 Suske, J., and Acker, G. 1989. Endophytic needle fungi: culture, ultrastructural and immunocytochemical studies. Pages 121–136 in: Ecological Studies, vol. 77. E.-D. Schulze, O. L. Lange and R. Oren, eds. Springer Verlag, Berlin, Heidelberg.
483 Sweigard, J. A., Chumley, F. G., and Valent, B. 1992. Disruption of a *Magnaporthe grisea* cutinase gene. Mol. Gen. Genet. 232:183–190.
484 Takabe, K., Fukazawa, K., and Harada, H. 1989. Deposition of cell wall components in conifer tracheids. Pages 47–66 in: Plant Cell Wall Polymers. Biogenesis and Biodegradation. N.G. Lewis and M.G. Paice, eds. ACS Symposium Series 399.
485 Takanori, I., and Terashima, N. 1990. Determination of the distribution and reaction of polysaccharides in wood cell walls by the Isotope Tracer Technique. Part I. Mokuzai Gakkaishi 36:917–922.
486 Taylor, J., Jones, J. D. G., Sandler, S., Mueller, G. M., Bedbrook, J., and Dunsmuir, P. 1987. Optimising the expression of chimeric genes in plant cells. Mol. Gen. Genet. 210:572–577.
487 Thiery, J. P. 1967. Mise en évidence des polysaccharides sur coupes fines en microscopie électronique. J. Microsc. (Paris) 6:987–1018.
488 Timell, T. E. 1965. Wood hemicelluloses. Part I. Adv. Carb. Chem. 19:247–302.
489 Timell, T. E. 1966. Wood hemicelluloses. Part II. Adv. Carb. Chem. 20:410–484.
490 Timell, T. E. 1986. Compression wood in Gymnosperms. Vol. 1. Springer–Verlag, Berlin–Heidelberg.
491 Tippett, J. T., and Shigo, A. L. 1981. Barriers to decay in conifer roots. Eur. J. For. Pathol. 11:51–59.
492 Tippett, J. T., Bogle, A. L., and Shigo, A. L. 1982. Response of balsam fir and hemlock roots to injuries. Eur. J. For. Path. 12:357–364.
493 Todd, D. 1988. The effects of host genotype, growth rate, and needle age on the distribution of a mutualistic, endophytic fungus in Douglas-fir plantations. Can. J. For. Res. 18:601–605.
494 Toti, L. 1993. The symbiosis *Discula umbrinella / Fagus sylvatica*: biochemical, ecological and morphological studies of the host endophyte relationship. Ph.D. Thesis ETH Zürich Nr. 10097, Switzerland.
495 Toti, L., Viret, O., Chapela, I. H., and Petrini, O. (1992). Differential attachment by conidia of the endophyte, *Discula umbrinella* (Berk. and Br.) Morelet, to host and non-host surfaces. New Phytol. 121:469–475.
496 Ulhoa, C. J., and Peberdy, J. F. 1991. Regulation of chitinase synthesis in *Trichoderma harzianum*. J. Gen. Microbiol. 137:2163–2169.
497 van den Ende, G., and Linskens, H.F. 1974. Cutinolytic enzymes in relation to plant pathogenesis. Annu. Rev. Phytopathol. 12:247–258.
498 van Kan, J. A. L., Cornelissen, B. J. C., and Bol, J. F. 1988. A virus-inducible tobacco gene encoding a glycine-rich protein shares regulatory element with the ribulose bisphosphate carboxylase small subunit gene. Mol. Plant-Microbe Interact. 1:107-112.
499 Vance, C. P., Kirk, T. K., and Sherwood, R. T. 1980. Lignification as a mechanism of disease resistance. Annu. Rev. Phytopathol. 18:259–288.
500 Vandenbosch, K. A. 1991. Immunogold labeling. Pages 181-219 in: Electron Microscopy of Plant Cells. J. L. Hall and C. Hawes, eds. Academic Press, New York.
501 Vanderhart, D. L., and Atalla, R.H. 1984. Studies of microstructures in

native celluloses using solid–state ^{13}C nmr. Macromolecules 17:1465–1472.
502 Vasseur, V., Arigoni, F., Andersen, H., Defago, G., Bompeix, G., and Seng, J. 1990. Isolation and characterization of *Aphanocladium album* chitinase over producing mutants. J. Gen. Microbiol. 136:2561–2567.
503 Vian, B., Brillouet, J. M., and Satiat-Jeunemaître, B. 1983. Ultrastructural visualization of xylans in cell walls of hard wood by means of xylanase-gold complex. Biol. Cell 49:179–182.
504 Vian, B., Nairn, J., and Reid, J. S. G. 1991. Enzyme-gold cytochemistry of seed xyloglucans using two xyloglucan-specific hydrolases. Importance of prior heat-deactivation of the enzymes. Histochem J. 23:116–124.
505 Vian, B., Reis, D., Mosiniak, M., and Roland, J. C. 1986. The glucuronoxylans and the helicoïdal shift in cellulose microfibrils in Linden Wood. Cytochemistry *in muro* and on isolated molecules. Protoplasma 131:185–199.
506 Viret, O. 1993. Infection of beech leaves by the endophyte *Discula umbrinella* (Berk. et Br.) Morelet [teleomorph: *Apiognomonia errabunda* (Rob.) Höhn.]: an ultrastructural study. Dissertation ETH No. 10073, ETH Zurich, Switzerland.
507 Viret, O., and Petrini, O. 1994. Penetration and colonisation of beech leaves (*Fagus sylvatica*) by the endophyte *Discula umbrinella* (Teleomorph: *Apiognomonia errabunda*). Mycol. Res., in press.
508 Viret, O., Scheidegger, C., and Petrini, O. 1993. Infection of beech leaves (*Fagus sylvatica*) by the endophyte *Discula umbrinella* (Teleomorph: *Apiognomonia errabunda*). Low temperature scanning electron microscopy studies. Can. J. Bot. 71:1520–1527.
509 Walker, C. 1992. Systematics and taxonomy of the arbuscular endomycorrhizal fungi (Glomales) – a possible way forward. Agronomie 12:887–897.
510 Wariishi, H., Valli, K., and Gold, M. H. 1991. In vitro depolymerizations of lignin by manganese peroxidase of *Phanerochaete chrysosporium*. Biochem. Biophys. Res. Commun. 176:269–275.
511 Watanabe, Y., Kaizu, S., and Koshijïma, T. 1986. Binding sites of carbohydrate moieties towards lignin in LCC from *Pinus densilora* wood. Chem. Lett. 11:1871–1874.
512 Weber, K., Rathke, P. C., and Osborne, M. (1978). Cytoplasmic microtubular images in glutaraldehyde-fixed tissue cells by electron microscopy and by immunofluorescence microscopy. Proc. Natl. Acad. Sci. USA 75:1820.
513 Westermark, U., and Eriksson, K. E. 1975. Purification and properties of cellobiose: quinone oxidoreductase from *Sporotrichum pulverulentum*. Acta Chem. Scand. B29:419–424.
514 Westermark, U., Lidbrandt, O., and Eriksson, I. 1988. Lignin distribution in spruce (*Picea abies*) determined by mercurization with SEM–EDXA technique. Wood Sci. Technol. 22:243–250.
515 White, J. F. 1988. Endophyte-Host associations in forage grasses. XI. A proposal concerning origin and evolution. Mycologia 80:442–446.
516 White, J. F., Angela, J., and Morrow, C. 1991. Endophyte-host associations in forage grasses. XIV. Primary stromata formation and seed transmission in *Epichloë typhina*: developmental and regulatory aspects. Mycologia 83:72–81.
517 White, J. F., Morrow, A. C., and Morgan-Jones, G. 1990. Endophyte-Host associations in forage grasses. XII. A fungal endophyte of *Trichachne insularis* belonging to *Pseudocercosporella*. Mycologia 82:218–226.

518 Wilkie, K. C. B. 1979. The hemicelluloses of grasses and cereals. Adv. Carb. Chem. Biochem. 36:215–264.
519 Wilkie, K. C. B., and Woo. S. L. 1977. A heteroxylan and hemicellulosic materials from Bamboo leaves. Carbohydr. Res. 57:145–162.
520 Wilson, D. 1992. On endophyte-insect-plant interactions. Ph.D. Thesis, University of Oregon, Eugene, Oregon, USA.
521 Wilson, T. 1990. Confocal microscopy. Academic Press, New York.
522 Woloshuk, C. P., and Kolattukudy, P. E. 1986. Mechanism by which contact with plant cuticle triggers cutinase gene expression in the spores of *Fusarium solani* f. sp. *pisi*. Proc. Natl. Acad. Sci. USA 83:1704–1708.
523 Wood, T. M. 1990. Fungal cellulases. Pages 491-533 in: Biosynthesis and Biodegradation of Cellulose and Cellulosic Materials. Weimer P. J. and Haigler C. A., eds. Marcel Dekker, New York.
524 Wood, T. M., and Garcia–Compayo, V. 1990. Enzymology of cellulose degradation. Biodegradation 1:147–161.
525 Wynn, W. K., and Staples, R. C. 1981. Tropisms of fungi in host recognition. Pages 45–69 in: Plant Disease Control: Resistance and Susceptibility. R. C. Staples and C. H. Toenniessen, eds. John Wiley and Sons, New York.
526 Yanai, K., Takaya, N., Kojima, N., Horiuchi, H., Ohtu, A., and Takagi, M. 1992. Purification of two chitinases from *Rhizopus oligosporus* and isolation and sequencing of the encoding genes. J. Bacteriol. 174:7398–7406.
527 Yang, Z., Cramer, C. L., and Watson, J. C. 1993. Protein farnesyltransferase in plants, molecular cloning and expression of a homologue of the β subunit from the garden pea. Plant Physiol. 101:667-674.
528 Zainal, A. S. 1976. The soft rot fungi: the effect of lignin. Mat. Org. Beih. 3:21–127.
529 Zarain–Herzberg, A., and Arroyo–Begovich, A. 1983. Chitinolytic activity from *Neurospora crassa*. J. Gen. Microbiol. 129:3319–3326.
530 Zhang, S., Sheng, J., Liu, Y., and Mehdy, M. C. 1993. Fungal elicitor-induced bean proline-rich mRNA down-regulation is due to destabilization that is transcription and translation dependent. Plant Cell 5:1089-1099.

DURHAM UNIVERSITY LIBRARY
3 0104 00772155 2